PHYSICS PRACTICAL
SCHEME OF WORK
FOR USE WITH THE IB DIPLOMA PROGRAMME

MICHAEL J. DICKINSON

For Nadine, Elena and Jacob

Acknowledgements

First published 2012

ISBN-10:   1475125798
ISBN-13: 978 - 1475125795

This work has been developed independently from
and is not endorsed by the International Baccalaureate (IB)

Front cover:
Photograph: Laszlo Varro
Graphic design: Nicholai Go

Thanks to:
the International Baccalaureate Organization
for permission to reproduce its intellectual property

Special thanks to:
Brian Seve, for his original training and inspiration.
Trevor Wilson, for the hours spent on proofreading and amendment suggestions.
Brian Seve, Ringo Dingrando and Gary Piech for their contributions.

CONTENTS

Investigations

Appendices

PREFACE

My first exposure to the International Baccalaureate program was back in about 1995, when I moved overseas to teach at the Colegio Anglo Colombiano, Bogotá, Colombia. Before I left the UK though, while I visiting an IB school in Liverpool, the head of science there tried to instil in me just how important it was to understand the requirements of the Internally Assessed (IA) portion of the syllabus. He offered me all sorts of advice regarding the experiment side of the IB Physics course and I left that day with my first Practical Scheme of Work (PSOW) for an IB Physics course.

In the first 6 months of teaching at "The Anglo" in Bogotá, I was sent to Cali, in the South of Colombia, on an IB workshop – it was a level 1 workshop for rookies just like me and a large portion of the course was dedicated to IA practicals. It was run by Brian Seve, who was at the time the head of science at The English School in Bogotá. I remember vividly a large science lab set up with about a dozen or so experiments around the edge. Each experiment had an accompanying student help sheet and I spent a couple of afternoons working my way around the experiments so that I would have a larger repertoire of labs to perform with my students when I returned to Bogotá. I took Brian's 4/PSOW away with me which I used extensively for the following few years.

A couple of years into my Anglo contract, my first cohort of seniors sat their IB exams. They performed reasonably well on the externally examined part, but I remember feeling disappointed with my results for their internally assessed work. When the marks had been published, together with the examiner's comments, I found to my dismay that all my students had been marked down! It transpired that I had been giving "too much information" to my students. For example, if I had told them both the dependent and independent variables and then assessed the lab on the Design (D) criteria students could not receive a "complete", even though I had marked it as such. Similarly, if I had given students the outline of a results table and suggested to them which graph they should draw, students could not receive a "complete" for Data Collection and Processing (DCP). A similar thing happened to the third criteria, Conclusion and Evaluation (CE) if I had given too much guidance.

It wasn't until I had moved to Helsinki, Finland in 1999 that I finally felt that I understood how much (or how little) information was acceptable for these student lab help sheets. Another colleague, Trevor Wilson had been a Workshop Leader for Physics for a number of years and he critiqued each of the lab sheets in my PSOW noting which criteria we could assess if we used them. There was absolutely nothing wrong with Brian's lab sheets, it was just that I had to be a little more discerning about which ones were suitable to assess which criteria.

A few years later I became an IB Moderator for Physics IAs and I remember receiving a sample of student work and feeling very nervous about the grade that I was about to award. The teacher had awarded one of his students 47/48, but I noticed that the teacher had told the students both the independent and dependent variables, when he was assessing the "Design" criterion. For the experiments that were being assessed for the "Data Collection and Processing" criterion, the teacher had told his students what data to collect, had given students results tables with the column headings already complete (so that all they had to do was record their data) and had told them what graph to draw and how to analyse it. The conclusions and evaluations of the labs were in the form of answers to a series of questions that the teacher had provided. I moderated the work down to 3/48, as the sample was not a true assessment of the capability of the student, but was more an indication of how well the student could follow instructions that the teacher had provided.

It should be noted that if the moderator finds a teacher's marking accurate, then no moderation factor is assigned to that teacher. However, if the moderator feels that the teacher has been a little too generous in their marking, then a moderation factor is applied to that teacher and the students in that teacher's classes may be marked down. Of course, if the teacher has been too harsh in their marking then the moderation factor applied to the teacher will benefit the students by increasing their overall IA mark.

I hope that you will find the information contained within this PSOW guide helpful and I wish you and your students every success in the internally assessed part of the IB Physics course.

Mike Dickinson – March 2012

INTERNAL ASSESSMENT

Practical scheme of work
Part of the requirement of the IB Syllabus for any of the experimental sciences in Group 4, is the completion of a "Practical Scheme of Work" or PSOW. This course of experimental study forms a very large part of a teacher's commitment and responsibility to ensure that students have been well prepared for experimental scientific endeavour, should they choose to pursue this type of career for their future. It is a responsibility since up to 24% of a student's overall grade is gained depending on how well they have performed by the end of this practical scheme of work. The teacher must be committed since there is a large time consideration to be made when designing the learning plan for the two year course – standard level students must complete a minimum of 40 hours of practical work, while higher level students must complete at least 60 hours (this time includes a maximum of 10 hours spent on the Group 4 Project but excludes the time spent writing up the work). Students in standard and higher level may perform some of the same investigations – this is particularly useful when, in smaller schools, SL and HL classes are often combined. Only 2 to 3 hours of investigative work can be carried out after the deadline for submitting work to the moderator and still be counted in the total number of hours for the practical scheme of work.

You don't have to assess every piece of practical work to the IB Internal Assessment criteria. There are often times when students use pieces of apparatus to support their learning and yet, the work is not going to be assessed. The hours spent on informal, non-assessed practical work can also be included in the 40 or 60 hours.

Syllabus Coverage
When designing the practical scheme of work, teachers should consider both breadth and depth of the syllabus being covered. It may be tempting, for example, to perform many experiments which cover the Mechanics and Electricity topics at the expense of covering the sections within, say, the Nuclear Physics topic. This is only natural as some topics lend themselves to a wider range of experimental study than other, but the teacher should attempt to spread the investigations over as many topics as possible. However, it is NOT mandated that students should carry out investigations for every syllabus topic.

A minimum number of investigations is not specified by the IBO, so teachers are quite at liberty to perform many smaller experiments or fewer, more in depth investigations, as they see fit. My own preference is to do a mixture of the two – in this way, the requirement that students undertake some "Design" (D) labs, which often take multiple lessons to complete, can be met. At the same time, I have included many examples of investigations that assess a student's ability to collect, manipulate and present data, so fulfilling the IB requirement of the "Data Collection and Presentation (DCP) criterion of the PSOW. The third assessment criterion is "Conclusion and Evaluation" (CE) and again, there are numerous opportunities within the included guide to assess this. Most of the lab sheets contained within this guide can also be used to assess a student's "Manipulative Skills" (MS). Some of them contain apparatus which can be quite trick to set up and manipulate in order to obtain good data – the Current Balance for example is particularly difficult to assemble, but once working the students can obtain some great data to verify the equation.

The Group 4 Project
One investigation that MUST be completed by ALL students though is the Group 4 Project. The Group 4 Project is an opportunity for students across the range of scientific disciplines to work together in order to investigate a topic from a range of angles. This interdisciplinary project allows Chemistry, Biology and Physics students (together with Computer Science, Environmental Systems & Societies, Sports Exercise & Health Science and Design Technology – if offered at your school) to engage in a collaborative project. The emphasis of this project is on process rather than product and so, the Group 4 Project assesses students "Personal Skill" (PS). I have included an example of the Group 4 Project that I undertook a few years ago while teaching in Hanoi. The overall theme of the project was "Science on and around the UNIS Hanoi, Tay Ho Campus" – student then investigated a wide variety of sub-topics within this general theme. Each of the small groups contained a physicist, a biologist and chemist and while each student within the e group was responsible for their particular science discipline, the whole group collaborated together to make these different aspects of physics to meld together into a coherent final report.

Acceptable Levels of Teacher Assistance
At first, it appears that the lab sheets provide very little information to the students. This is by design rather than by accident. As I said in the preface, it is important that the lab reports submitted by the students are their own work and not the work of an experienced teacher. Here are some dos and don'ts when it comes to the amount of assistance that you can give to students.

When assessing the Design (D) criterion.
- ✓ It is acceptable to give students a general theme to investigate.
- ✓ It is acceptable to give the students the dependent variable.
- ✗ It is unacceptable to give students both the dependent and independent variables to investigate.
- ✗ It is unacceptable to give students the controlled variables
- ✓ It is acceptable to show them the range of equipment that your school has for the topic being investigated.
- ✗ It is unacceptable to give students a materials and apparatus list.
- ✓ It is acceptable to explain to students how a piece of apparatus works.
- ✗ It is unacceptable to give students a method for the collection of data.
- ✓ It is acceptable to teach students how to propagate errors in experiments.
- ✗ It is unacceptable to explain how variables should be controlled (for that specific experiment).

When assessing the Data Collection and Processing (DCP) criterion
- ✓ It is acceptable to suggest which variables to collect data for (as long as the experiment is not being assessed for the Design criterion also).
- ✗ It is unacceptable to provide a blank table with headings and units, which is simply to be filled in by the student(s).
- ✓ It is acceptable to suggest that the student(s) process the data in a "suitable" way.
- ✗ It is unacceptable to tell the students to plot a graph of variable A vs. variable B.
- ✗ It is unacceptable to provide pre-labelled graph axes on which students need only to plot points.
- ✓ It is acceptable to suggest that the student(s) investigate the degree of uncertainty in their calculated result
- ✗ It is unacceptable to tell the students exactly how to process the result in order to calculate the uncertainty – they should have already been taught this prior to the investigation.
- ✗ It is unacceptable to provide students with a set of structured questions to be answered.

When assessing the Conclusion and Evaluation (CE) criterion
- ✓ It is acceptable to provide minimal guidance on how a conclusion might be structured.
- ✓ It is acceptable to suggest that students compare the result obtained in their experiment to that quoted in the textbook.
- ✗ It is unacceptable to provide students with a set of structured questions to be answered.

As part of the learning process, teachers can give general advice to students on a first draft of their work for IA. However, constant drafting and redrafting is not allowed and the next version handed to the teacher after the first draft must be the final one. In assessing student work using the IA criteria, teachers should only mark and annotate the final draft. This is marked by the teacher using the IA criteria. It is useful to annotate this work with the levels awarded for each aspect—"c" for complete, "p" for partial and "n" for not at all. This will both provide useful feedback to the student and will assist the moderator should the work be selected as part of the sample to be sent for moderation. [from the IB Physics Diploma Programme Guide © International Baccalaureate Organisation 2007]

Types of Investigations
The IA model is flexible enough to allow a wide variety of investigations to be carried out. These could include:
- short laboratory practicals over one or two lessons and long-term practicals or projects extending over several weeks.
- computer simulations.
- data-gathering exercises such as questionnaires, user trials and surveys.
- data-analysis exercises.
- general laboratory work and fieldwork.
  [from the IB Physics Diploma Programme Guide © International Baccalaureate Organisation 2007]

Use of Information and Communication Technology (ICT)
One of the aims of IB is to "develop and apply the students' information and communication technology skills in the study of science". The use of information and communication technology (ICT) is encouraged in practical work throughout the course, whether the investigations are assessed using the IA criteria or otherwise.

It is not necessary to use ICT in all assessed investigations but, in order to carry out aim 7 in practice, students will be required to use each of the following software applications at least once during the course.
- Data logging in an experiment
- Software for graph plotting
- A spreadsheet for data processing
- A database
- Computer modelling/simulation

Apart from sensors for data logging, all the other components involve software that is free and readily available on the Internet. As students only need to use data-logging software and sensors once in the course, class sets are not required. The use of each of the above five ICT applications by students would be authenticated by means of entries in the students' practical scheme of work, form 4/PSOW. For example, if a student used a spreadsheet in an investigation, this should be recorded on form 4/PSOW. Any other applications of ICT can also be recorded on form 4/PSOW. [from the IB Physics Diploma Programme Guide © International Baccalaureate Organisation 2007]

The Form 4/PSOW
The results of each student's assessment for each investigation that they carry out as they progress through the Practical Scheme of Work should be recorded on the Form 4/PSOW. The IB provide the 4/PSOW form in section 4 of the "Handbook of procedures for the Diploma Programme" – this used to be called the "Vade Mecum" until IB realized that nobody understood what "Vade Mecum" meant. On pages xiii and xiv, I have provided my own version of this form as an example of how it should be filled out. The following MUST be included on the form:

At the top of the form:
- Submit to   Moderator
- Arrival date   April / October (depending on the session of the exam)
- Session   May / November
- School number   00 XXXX
- School name   Use words rather than abbreviations
- Subject   Physics
- Level   Standard / Higher level
- Candidate name   Use the name provided to IB when the student registered
- Candidate number   00 XXXX XXX

In the body of the form:
- Date(s)   The date that the investigation was carried out
- Outline of experiment / investigation   This doesn't have to be too detailed
- ICT   Number 1 – 5 corresponding to the ICT type
- Topic / option   Topic and sub-topic from the syllabus guide
- Time (hrs)   I used 0.5 hours as a minimum
- Levels awarded   The level reached (max 6) on the 5 different marking criteria

At the end of the form:
- Two highest levels achieved   Not an average, or a best guess – must be the two highest
- Total / 48   This is described on the next page

- Teacher's name, Teacher's signature and date
- Candidate's signature and date

Form 4/IA

In addition to the Form 4/PSOW (one form for each student), teachers will also need to complete the Form 4/IA. One form is needed to accompany the complete package of (usually) 10 samples which are sent for moderation. The form is a checklist to ensure that teachers have done the following:

- read section A10.6 and A10.7 and section 4 in the handbook.
- participated in internal standardization where two or more teachers are responsible for the sample material.
- included a form 4/PSOW for each candidate in the sample set, signed and dated by the teacher.
- checked that photocopied material is legible (ideally, original work should be sent to the moderator).
- made sure that the criteria D, DCP and CE, have all been assessed on at least two occasions.
- highlighted or circled the two highest levels for each of the criteria D, DCP and CE on each PSOW form.
- included the corresponding write-ups/reports and teacher instruction sheets for each candidate.
- included the title of the group 4 project is on the 4/PSOW and noted the level achieved for PS
- noted the summative mark for MS.
- flagged the experiments/dates on which the students experienced specific ICT applications.

Calculating the students IA mark

A maximum of 24% of the overall IB score is gained by students completing their "Practical Scheme of Work (PSOW)". For Higher Level students, this PSOW amounts to 60 hours of practical laboratory lab work and investigations. For Standard Level students, the time allocation is 40 hours.

There are five assessment criteria that are used to assess the work of both SL and HL students;
- Design (D)
- Data collection and processing (DCP)
- Conclusion and evaluation (CE)
- Manipulative skills (MS)
- Personal skills (PS)

Each time a student is assessed for one or more of the criteria, they can be awarded a maximum of 6 points for that criterion. The "c, p, n" notation is used to assess the work:

    c = complete = 2 points,
    p = partial   = 1 point
    n = not at all = 0 points.

By being awarded "complete" (c, c, c) for all three aspects of a particular criterion, the student will receive the maximum 6 points. (c, c, p) will receive 5 points, (c, p, p) = 4 points .....etc.

The first three criteria – design (D), data collection and processing (DCP) and conclusion and evaluation (CE) – are each assessed twice. Manipulative skills (MS) is assessed summatively over the whole course and the assessment should be based on a wide range of manipulative skills. Personal skills (PS) is assessed only once, during the "Group 4 Project". The table below shows how the students mark out of 24 is calculated.

Design (D)	2 x 6	= 12	(best 2 levels achieved by the students)
Data collection and processing (DCP)	2 x 6	= 12	(best 2 levels achieved by the students)
Conclusion and evaluation (CE)	2 x 6	= 12	(best 2 levels achieved by the students)
Manipulative skills (MS)	1 x 6	= 6	(assessed summatively over the whole course)
Personal skills (PS)	1 x 6	= 6	(assessed during the Group 4 Project)
Total / 48		= 48	($\div$ 2 = the students' final score out of 24)

Marking Rubrics

Pages xi and xii contain reproductions of the marking rubrics supplied by IBO in the Physics syllabus guide – first examinations 2009. These rubrics are to be used by the teacher to establish the level (c, p or n) of each the 3 aspect on each of the 5 marking criteria. On page xv I have included the form that I use for the back side of each of the lab sheets – it only contains the first 3 criteria. It is important that students understand how they will be assessed so that, over the two years of the course, with more and more exposure to this rubric they can improve.

Sample Lab Report

Pages xvi to xxiii contains an exemplar of an investigation that has been used to assess students on the first 3 of the assessment criteria above – design (D), data collection and processing (DCP) and conclusion and evaluation (CE). Immediately following this exemplar, I have then added a section containing the same piece of work, including supporting notes so that the detail and complexity of what is required to achieve "complete (c)" in all aspects can be seen.

INTERNAL ASSESSMENT MARKING RUBRICS

Criterion 1 – Design (D)

Levels/marks	Aspect 1 — Defining the problem and selecting variables	Aspect 2 — Controlling variables	Aspect 3 — Developing a method for collection of data
Complete/2	Formulates a focused problem/research question and identifies the relevant variables.	Designs a method for the effective control of the variables.	Develops a method that allows for the collection of sufficient relevant data.
Partial/1	Formulates a problem/research question that is incomplete **or** identifies only some relevant variables.	Designs a method that makes some attempt to control the variables.	Develops a method that allows for the collection of insufficient relevant data.
Not at all/0	Does not identify a problem/research question **and** does not identify any relevant variables.	Designs a method that does not control the variables.	Develops a method that does not allow for any relevant data to be collected.

Criterion 2 – Data Processing and Presentation (DPP)

Levels/marks	Aspect 1 — Defining the problem and selecting variables	Aspect 2 — Controlling variables	Aspect 3 — Developing a method for collection of data
Complete/2	Records appropriate quantitative and associated qualitative raw data, including units and uncertainties where relevant.	Processes the quantitative raw data correctly.	Presents processed data appropriately and, where relevant, includes errors and uncertainties.
Partial/1	Records appropriate quantitative and associated qualitative raw data, but with some mistakes or omissions.	Processes quantitative raw data, but with some mistakes and/or omissions.	Presents processed data appropriately, but with some mistakes and/or omissions.
Not at all/0	Does not record any appropriate quantitative raw data **or** raw data is incomprehensible.	No processing of quantitative raw data is carried out **or** major mistakes are made in processing.	Presents processed data inappropriately **or** incomprehensibly.

Criterion 3 – Conclusion and Evaluation (CE)

Levels/marks	Aspect 1 — Defining the problem and selecting variables	Aspect 2 — Controlling variables	Aspect 3 — Developing a method for collection of data
Complete/2	States a conclusion, with justification, based on a reasonable interpretation of the data.	Evaluates weaknesses and limitations.	Suggests realistic improvements in respect of identified weaknesses and limitations.
Partial/1	States a conclusion based on a reasonable interpretation of the data.	Identifies some weaknesses and limitations, but the evaluation is weak or missing.	Suggests only superficial improvements.
Not at all/0	States no conclusion **or** the conclusion is based on an unreasonable interpretation of the data.	Identifies irrelevant weaknesses and limitations.	Suggests unrealistic improvements.

Criterion 4 – Manipulative Skills (MS)

Levels/marks	Aspect 1 Defining the problem and selecting variables	Aspect 2 Controlling variables	Aspect 3 Developing a method for collection of data
Complete/2	Follows instructions accurately, adapting to new circumstances (seeking assistance when required).	Competent and methodical in the use of a range of techniques and equipment.	Pays attention to safety issues.
Partial/1	Follows instructions but requires assistance.	Usually competent and methodical in the use of a range of techniques and equipment.	Usually pays attention to safety issues.
Not at all/0	Rarely follows instructions **or** requires constant supervision.	Rarely competent and methodical in the use of a range of techniques and equipment.	Rarely pays attention to safety issues.

Criterion 5 – Personal Skills (PS)

Levels/marks	Aspect 1 Defining the problem and selecting variables	Aspect 2 Controlling variables	Aspect 3 Developing a method for collection of data
Complete/2	Approaches the project with self-motivation and follows it through to completion.	Collaborates and communicates in a group situation and integrates the views of others.	Shows a thorough awareness of their own strengths and weaknesses and gives thoughtful consideration to their learning experience.
Partial/1	Completes the project but sometimes lacks self-motivation.	Exchanges some views but requires guidance to collaborate with others.	Shows limited awareness of their own strengths and weaknesses and gives some consideration to their learning experience.
Not at all/0	Lacks perseverance and motivation.	Makes little or no attempt to collaborate in a group situation.	Shows no awareness of their own strengths and weaknesses and gives no consideration to their learning experience.

[from the IB Physics Diploma Programme Guide © International Baccalaureate Organisation 2007]

SAMPLE FORM 4/PSOW

Submit to: __Moderator__   Arrival date: __April 2012__   Session: __May 2012__

School code: __00 0001__   School name: __International School of South East Asia__

Subject: __Physics__   Level: __Higher Level__

Candidate name: __Bond, James__   Candidate number: __00001 007__

Date(s)	Outline of experiments / investigations / projects (include title and a brief description)	ICT	Topic / option	Time (hrs)	Levels awarded		
					D	DCP	CE
Aug '10	Investigating Murphy's Law		----	1.0	0	1	0
Aug '10	Investigating the Fall of a Coffee Filter	2,3	1	2.0	1	1	0
Aug '10	Investigating the Simple Pendulum	2,3	1	1.0	1	2	1
Sep '10	Investigating Uncertainties – Measuring Instrument Circus		1	1.0		2	
----------	Investigating Errors and Uncertainties in Experiments		----	---		X	
Sep '10	Determining Stiffness of Steel by the Oscillations of a Hacksaw Blade	1,2,3	1	1.0		2	1
Oct '10	Investigating the Stopping Distance of a Bicycle	2,3	1, 2	3.0	1	2	2
Oct '10	Investigating the Torsional Pendulum	2,3	2	1.0	2	2	2
Oct '10	Investigating Projectiles	2,3	2	1.0		3	4
Nov '10	Investigating Forces in Equilibrium		2	1.0		3	
----------	Investigating Error and uncertainties using Acceleration due to gravity		----	----		X	X
Nov '10	Investigating the Flight of an Elastic Band	2,3,5	2, 9	3.0	4	2	3
Dec '10	Investigating Work Done and Energy Transferred on an Inclined Plane		2	1.0		3	3
Dec '10	Investigating the Ballistic Pendulum		2, 9	1.0		3	4
----------	Investigating Energy Transfer and Energy Loss of a Rolling Ball		----	----		X	
Jan '11	Investigating Newton's 2nd Law of Motion using a Ticker timer	2,3	2	2.0		4	4
Jan '11	Investigating Hooke's Law		2	1.0		3	4
Feb '11	Investigating Springs in Series and Parallel		2	2.0	2	3	3
Feb '11	Investigating Circular Motion	2,3	2	1.0		4	3
Feb '11	Investigating Ping - Pong Balls	2,3	3	2.0	3	3	4
Mar '11	Determining Specific Heat Capacity by The Electrical Method	1,2,3	3	1.0		5	4
Mar '11	Determining Specific Heat Capacity by The Method of Mixtures		3	1.0		4	4
----------	Investigating Specific Latent Heat of Vaporisation of Water	2,3	----	----		X	X
Apr '11	Investigating Specific Latent heat of Fusion of Ice		3	1.0		5	5
Apr '11	Investigating the Power and Temperature of the Sun		3	1.0		3	4
Apr '11	Investigating Charles' Law and Absolute Zero		3	1.0			4
May '11	Investigating Rate of Cooling	1,2,3,5	3	1.0		4	5

May '11	Investigating the Pressure / Volume relationship for a Balloon	2,3	3	2.0		4	
----------	Determining the Temperature of a Wire by Expansivity		----	----		X	
Aug '11	Determining the Refractive Index of Glass by Real and Apparent Depth		4	1.0		5	
Aug '11	Investigating Refraction of Light		4	1.0		3	
Aug '11	Investigating Malus' Law		4, 11	1.0		4	4
Sep '11	Investigating Brewster's Law		11	1.0		5	5
----------	Investigating the Focal Length of a Converging Lens		----	----		X	X
Sep '11	Investigating Melde's Experiment		11	1.0		5	5
Oct '11	Investigating Resonance – Determining the Velocity of Sound		11	1.0		4	4
Oct '11	Determining the Wavelength of Laser Light using Young's Double Slits	4	11	1.0		4	4
Oct '11	Determining the Wavelength of light using a Diffraction Grating	4	11	1.0		5	5
Nov '11	Investigating the Power of an Electric Heater		5	1.0		5	
----------	Investigating Resistance Wire		----	----	X	X	X
Nov '11	Determining Energy Density of Fuels	4	8	1.0		5	5
Dec '11	Investigating Lenz's Law using the Motion of a Falling Magnet		9	1.0		4	5
Dec '11	Investigating Magnets		9	2.0	4	4	5
Jan '12	Verifying the Equation "F = B I L" Using a Current Balance		9, 11	1.0		5	5
Jan '12	Investigating Electromagnets		12	2.0	⑤	⑤	⑥
----------	Investigating the Efficiency of a Transformer		----	----		X	
Feb '12	Investigating Electromagnetic Induction		12	2.0		4	4
Feb '12	Determining the Emf and Internal Resistance of a cell		12	1.0		⑤	5
Mar '12	Investigating Radioactive Decay (Simulation using Dice)	5	7, 13	1.0	⑤	4	⑤
Jan '11	Group 4 Project – An investigation into the energy efficiency of Systems on the ISM campus – Mark for Personal Skills (PS)	1,2,3,4	8	10			

Group 4 Project mark for Personal Skills (PS) (same mark for students doing 2 subjects)	4 /6	Two highest levels achieved	5 /6 5 /6 6 /6
Summative Mark for Manipulative Skills (MS)	3 /6		5 /6 5 /6 5 /6

For completion by the examiners			Total
Moderator /6 /6 /6 /6 /6 /6		Senior Moderator /6 /6 /6 /6 /6 /6	38 /48 This total must also be entered on IBIS

To be completed by the teacher:

Name: **Michael John Dickinson**　　　　Signature: _____　　　　Date: **March 5th 2012**

Candidate declaration: I confirm that this work is my own work and is the final version. I have acknowledged each use of the words or ideas of another person, whether written, oral or visual.

Candidate's signature: _____　　　　Date: _____

INTERNAL ASSESSMENT MARKING FORM

Criteria		Aspect 1	Aspect 2	Aspect 3	Level awarded
Design (D)		**Defining the Problem and selecting variables:**	**Controlling the Variables:**	**Developing a method for collecting data:**	
	c = 2	Formulates a focused problem/research question and identifies the relevant variables.	Designs a method for the effective control of the variables.	Develops a method that allows for the collection of sufficient relevant data.	
	p = 1	Formulates a problem/research question that is incomplete **or** identifies only some relevant variables.	Designs a method that makes some attempt to control the variables.	Develops a method that allows for the collection of insufficient relevant data.	
	n = 0	Does not identify a problem/research question **and** does not identify any relevant variables.	Designs a method that does not control the variables.	Develops a method that does not allow for any relevant data to be collected.	/6
Data Collection and Processing (DCP)		**Recording raw data**	**Processing raw data**	**Presenting processed data**	
	c = 2	Records appropriate quantitative and associated qualitative raw data, including units and uncertainties where relevant.	Processes the quantitative raw data correctly.	Presents processed data appropriately and, where relevant, includes errors and uncertainties.	
	p = 1	Records appropriate quantitative and associated qualitative raw data, but with some mistakes or omissions.	Processes quantitative raw data, but with some mistakes and/or omissions.	Presents processed data appropriately, but with some mistakes and/or omissions.	
	n = 0	Does not record any appropriate quantitative raw data **or** raw data is incomprehensible.	No processing of quantitative raw data is carried out **or** major mistakes are made in processing.	Presents processed data inappropriately **or** incomprehensibly.	/6
Conclusion and Evaluation (CE)		**Concluding**	**Evaluating procedure(s)**	**Improving the investigation**	
	c = 2	States a conclusion, with justification, based on a reasonable interpretation of the data.	Evaluates weaknesses and limitations.	Suggests realistic improvements in respect of identified weaknesses and limitations.	
	p = 1	States a conclusion based on a reasonable interpretation of the data.	Identifies some weaknesses and limitations, but the evaluation is weak or missing.	Suggests only superficial improvements.	
	n = 0	States no conclusion or the conclusion is based on an unreasonable interpretation of the data.	Identifies irrelevant weaknesses and limitations.	Suggests unrealistic improvements.	/6

Sample Physics Laboratory Report (Investigating a Bouncing Spring)

Investigating a Bouncing Spring

Research Question:

The aim of this experiment is to investigate how the mass on the end of a spring affects the time period of oscillation.

Hypothesis:

I believe that as the mass on the end of the spring increases, the time period will increase. If we double the mass, then the period of the spring will double – in other words mass is directly proportional to time period.

I think that this will happen because Newton's 2nd law says that with twice the mass, the rate of acceleration will be halved, causing a lower average speed and therefore an increase in the time period for one swing.

I think that if a graph is plotted of time period against mass it will be a straight line graph indicating a directly proportional relationship.

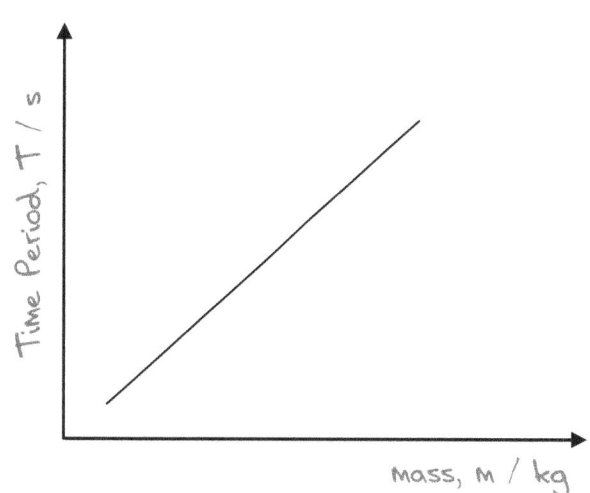

Variables List:

Independent variable   –   mass, m (kg)
Dependent variable   –   time period, T (s)
Controlled variables   –   length of spring, l (m)
                          –   spring constant, k (Nm⁻¹)
                          –   amplitude of oscillation, A (m)
                          –   friction (air resistance), F (N)
                          –   acceleration due to gravity, g (ms⁻²)

Controlling the Variables:

Mass, m
The mass on the end of the spring will be changed from 100g to 1 kg, 100g at a time.
Time period, T
The time period of the oscillating spring will be measured using a digital stopwatch. The spring will be timed for 10 oscillations and this time will be divided by 10 to improve accuracy. This procedure will be repeated 3 times and an average taken.
Length of spring, l
As more mass is added, the spring's extension will increase, however, the amplitude will be kept constant

Spring constant

The spring constant will remain constant each time the mass is changed. It will be controlled by using the same spring.

Amplitude of oscillation, A

The amplitude will remain constant each time the mass is changed. It will be controlled using a metre rule.

Friction (air resistance), F

Air resistance is difficult to control. However, it can be minimised by releasing the spring with small amplitude. The ceiling fans will be turned off during the experiment and the same location will be used throughout.

Acceleration due to gravity, g

As long as the experiment is performed at the same location then acceleration due to gravity will remain constant.

Materials List:

Retort Stand
Clamp and boss
Spring
Mass holder (100g)
10 x 100g slotted masses
Metre rule
Stopwatch

Diagram:

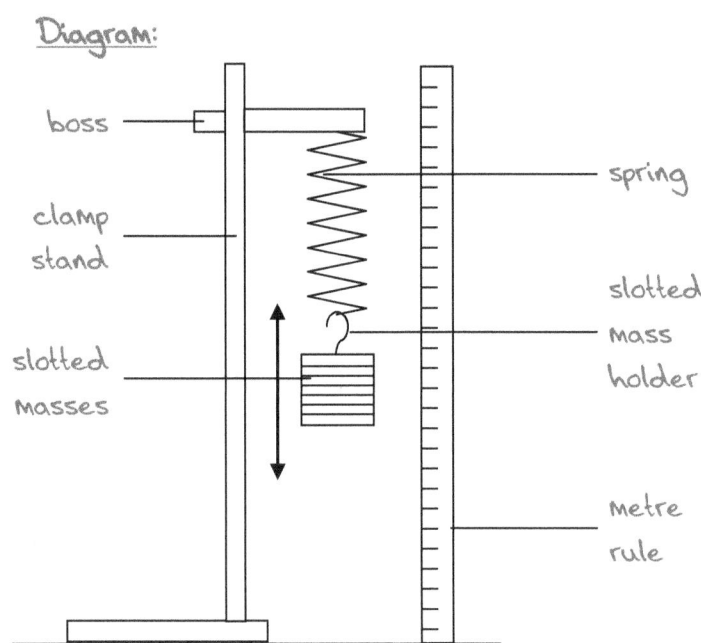

Method:

1. Set up the apparatus as shown in the diagram, with one end of the spring attached to the horizontal support on the clamp stand.
2. Attach the slotted mass holder to the bottom of the spring.
3. Position the metre rule next to (but not touching) the spring so that the amplitude of oscillation can be controlled.
4. Pull the 100g slotted mass holder down by 5cm, using a ruler and release the spring.
5. Time 10 complete oscillations of the spring using the stopwatch and record the data.
6. Repeat step 4 two more times (3 times in total).
7. Record the data each time and take an average.
8. Add 100g to the slotted mass holder and repeat steps 4 – 6 until the mass on the slotted mass holder is 1 kg

Results:

Table 1 - Mass on the spring and time period

Mass, m / kg	Δm / kg	Time for 10 oscillations of the spring / s ΔT = ±0.01 s				Time for 1 oscillation / s ΔT = ±0.01 s
		T_1 / s	T_2 / s	T_3 / s	Average time T_{AVE} / s	T / s
0.100	0.001	2.30	2.34	2.26	2.30	0.23
0.200	0.002	3.29	3.27	3.25	3.27	0.33
0.300	0.003	4.04	4.09	4.12	4.08	0.41
0.400	0.004	4.65	4.62	4.72	4.66	0.47
0.500	0.005	5.19	5.21	5.13	5.18	0.52
0.600	0.006	5.69	5.66	5.72	5.69	0.57
0.700	0.007	6.10	6.12	6.03	6.08	0.61
0.800	0.008	6.59	6.50	6.53	6.54	0.65
0.900	0.009	6.93	6.88	6.87	6.89	0.69
1.000	0.010	7.31	7.25	7.31	7.29	0.73

Data Presentation:

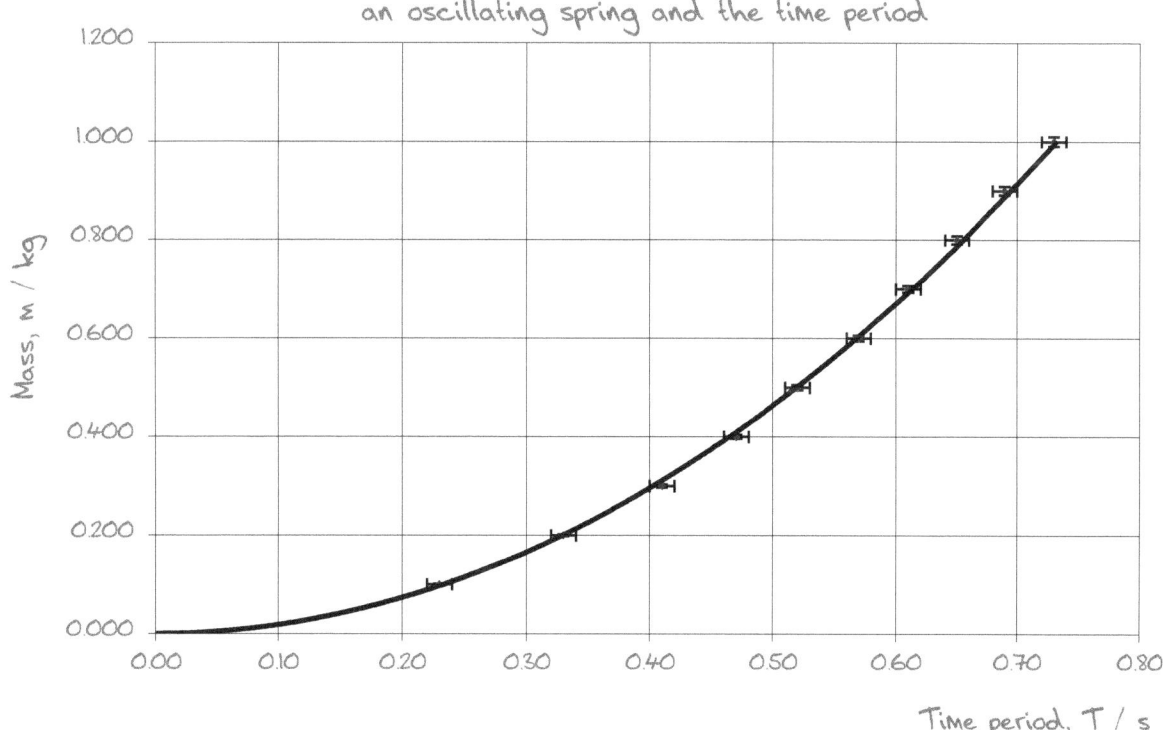

A graph to show the relationship between the mass on the end of an oscillating spring and the time period

Analysis of the Raw Data:

The relationship between mass and time period for an oscillating spring appears to be non-linear. The data will therefore be processed in order to find a relationship between these two variables. Since the graph appears to be parabolic, the data will be manipulated so that a graph of mass vs. time^2 can be plotted.

Data Processing:

Table 2 – Mass on the spring and time period squared, T^2.

Mass, m / kg $\Delta m = \pm 0.001$ kg	Time, T / s $\Delta T = \pm 0.01$ s	uncert. in T / %	Time^2, T^2 / s^2	uncert. in T^2 / %	uncert. in T^2 / s^2
0.100	0.23	4.35	0.053	8.70	0.0046
0.200	0.33	3.03	0.109	6.06	0.0066
0.300	0.41	2.44	0.168	4.88	0.0082
0.400	0.47	2.13	0.221	4.26	0.0094
0.500	0.52	1.92	0.270	3.84	0.0104
0.600	0.57	1.75	0.325	3.50	0.0114
0.700	0.61	1.64	0.372	3.28	0.0122
0.800	0.65	1.54	0.423	3.08	0.0130
0.900	0.69	1.45	0.476	2.90	0.0138
1.000	0.73	1.37	0.533	2.74	0.0146

Presentation and Analysis of the Processed Data:

A graph to show the relationship between the mass, m, on the end of an oscillating spring and the time period squared, T^2.

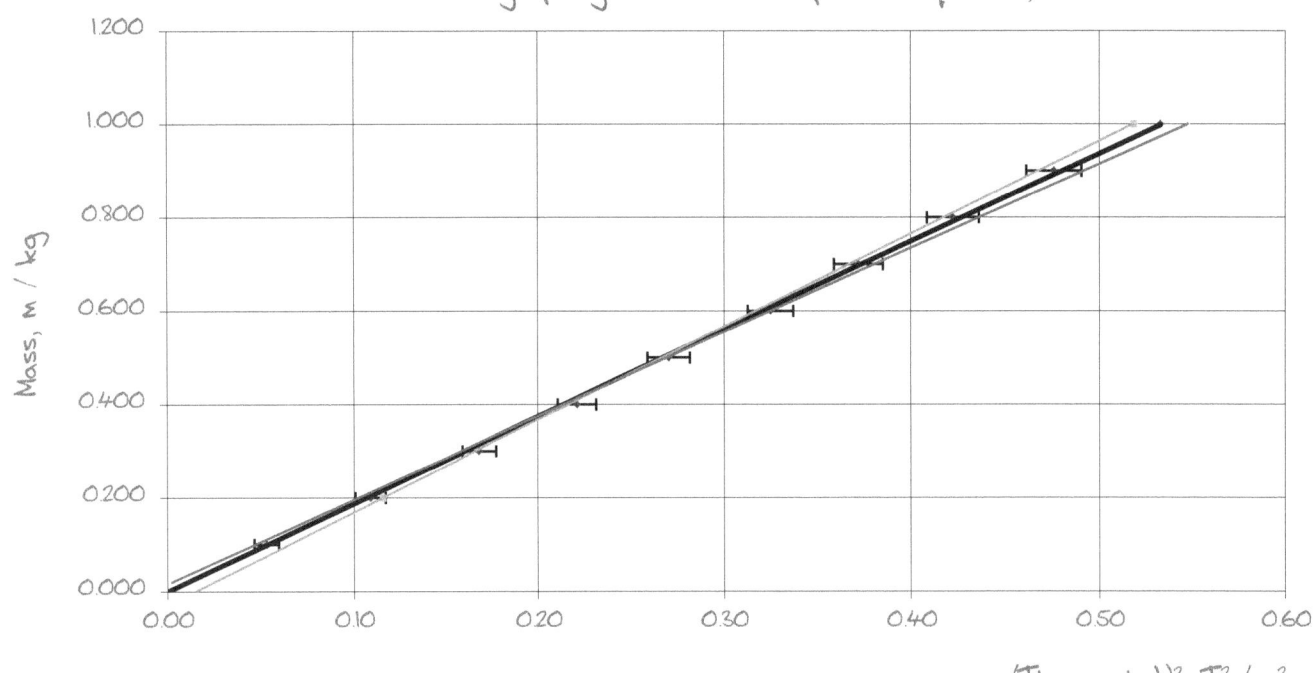

Analysis of the graph:

Gradient of best fit line $= \dfrac{y_2 - y_1}{x_2 - x_1} = \dfrac{1.00}{0.53} = 1.87\,\text{kgs}^{-2}$

Gradient of steepest line $= \dfrac{y_2 - y_1}{x_2 - x_1} = \dfrac{1.00}{0.51} = 1.99\,\text{kgs}^{-2}$

Gradient of shallowest line $= \dfrac{y_2 - y_1}{x_2 - x_1} = \dfrac{0.98}{0.54} = 1.80\,\text{kgs}^{-2}$

From this second graph it can be seen that the mass on the end of an oscillating spring is directly proportional to the time period squared.

Mathematically, $\qquad T^2 \propto M$

After investigation it was found that the formula relating the time period of an oscillating spring to the mass is

$$T = 2\pi\sqrt{\frac{M}{k}}$$

re-arranging this equation

$$T^2 = \frac{4\pi^2}{k} \cdot M$$

which is of the form $\qquad y = mx \qquad$ (equation of a straight line)

The gradient of the best fit line is therefore equal to $4\pi^2/k$, the spring constant can now be found:

$$gradient = 1.87 = \frac{k}{4\pi^2}$$
$$k = 1.87 \times 4\pi^2$$
$$k = 73.8 \text{ kgs}^{-2} \text{ (Nm}^{-1})$$

The range of uncertainty in this value can be calculated using the maximum and minimum lines on the graph

$$max \; gradient = 1.99 = \frac{k}{4\pi^2} \qquad\qquad min \; gradient = 1.80 = \frac{k}{4\pi^2}$$
$$k = 1.99 \times 4\pi^2 \qquad\qquad\qquad\qquad k = 1.80 \times 4\pi^2$$
$$k = 78.6 \text{ kgs}^{-2} \text{ (Nm}^{-1}) \qquad\qquad k = 71.1 \text{ kgs}^{-2} \text{ (Nm}^{-1})$$

therefore the spring constant, $k = 73.8 \, ^{+4.8}_{-2.7} \text{ Nm}^{-1}$

Conclusion:

The aim of this experiment was to investigate how the mass on the end of a spring affects the time period of oscillation. I predicted that the time period would be directly proportional to the mass of the oscillating system. This prediction turned out to be incorrect as the graph of mass vs. time period is clearly non-linear. The second graph of mass vs. time period squared however turned out to be linear and therefore I can conclude that mass is directly proportional to the time period squared.

After further investigation this conclusion is supported as the equation for the time period of an oscillating spring is

$$T = 2\pi\sqrt{\frac{M}{k}}$$

re-arranging this equation

$$T^2 = \frac{4\pi^2}{k} \cdot M$$

so

$$T^2 \propto M$$

The gradient of the straight line was then used to calculate the spring constant, k for the spring used in this experiment.

This was found to be k = $73.8 \, ^{+4.8}_{-2.7} \, Nm^{-1}$

This cannot be compared to a literature value, but a simple Hooke's Law experiment can be used to verify the validity of this result. When the spring was loaded to 10N, the extension was seen to be 14.2 cm (0.142 m). From this the spring constant can be determined.

$$F = kx$$

$$k = \frac{F}{x} = \frac{10}{0.142} = 70.4 \, Nm^{-1}$$

Comparing the two results, a percentage deviation can be calculated.

$$Percentage\ deviation = \frac{73.8 - 70.4}{73.8} = 4.6\%$$

This is an acceptable difference (within 5%) and therefore the experiment can be assumed to be successful and the results accurate.

Evaluation:

In general the method and apparatus worked well. There were however some modifications that were made when collecting the data that were not stated in the original plan.

1. Parallax error (random error) when reading the ruler was accounted for by placing the recorders eye level with the bottom of the spring.

2. When lighter loads were used (100g and 200g), the period of oscillation was so fast that it was quite difficult to count ten oscillations. In these cases, more than three trials were recorded and the then three were selected to use in the table of results based on how close they were to each other. This was a random error that caused uncertainty in the period.

3. The slotted masses were never checked for their accuracy. The 100g stamped on each was taken to be accurate. This may have produced a systematic error, depending on how inaccurate the masses were and the consistency of their inaccuracy

4. I found that the 1kg mass was a very large mass to use. It was so heavy that it caused the whole system, to bounce around uncontrollably. This was a random error that caused uncertainty in the period.

Suggested Improvements:

The investigation could have yielded more accurate results if the following modifications were made in future:

1. Parallax error. A horizontal pin could have been stuck to the bottom of the slotted mass holder. This would have pointed at the ruler so that the elimination of parallax error simply by guess would have been eliminated.

2. Timing issues relating to human reaction could have been avoided if the "Vernier" ultra-sonic motion detector was placed below the oscillating spring. The data would have therefore been collected automatically and human interaction would have been avoided.

3. A balance should have been used to check to accuracy of each of the 100g slotted masses.

4. For next time, the range of masses should be changed to 50g – 500g in increments of 50g.

SAMPLE PHYSICS LABORATORY REPORT
(WITH SUPPORTING NOTES)

SAMPLE PHYSICS LABORATORY REPORT (WITH SUPPORTING NOTES)

Design

Aspect 1 – Defining the Problem and selecting variables:

Try to think of a **focussed** research question in which one variable (**independent variable**) will affect another variable (**dependent variable**).

e.g.   *The aim of this experiment is to investigate how the mass on the end of a spring affects the time period of oscillation.*

It is a good idea at this stage to think about how you will process your raw data. A graph of the dependent variable (time in this case) vs. the independent variable (mass) will be plotted which will lead to a conclusion as to the relationship between them.

Hypothesis (this has become optional in the latest syllabus guide, but I still like students to write one)
The hypothesis should be **quantitative** (if possible) and should relate directly to your **research question**. The hypotheses should be **explained**.

e.g.   Qualitative
I believe that as the mass on the end of the spring increases, the time period will increase.

Quantitative
If we double the mass, then the period of the spring will double – in other words mass is directly proportional to time period.

Explained
I think that this will happen because Newton's 2nd law says that with twice the mass, the rate of acceleration will be halved, causing a lower average speed and therefore an increase in the time period for one swing.

A predictive graph is also a good way to support and clarify your hypothesis. It is also a good way to ensure that your hypothesis is quantitative.

e.g.   *I think that if a graph is plotted of time period against mass it will be a straight line graph indicating a directly proportional relationship.*

Note: It doesn't matter if the hypothesis is right or wrong! In this case, I've tried to write a hypothesis which is flawed scientifically, but still would receive full marks since it is quantitative, relates directly to the research question and can be tested.

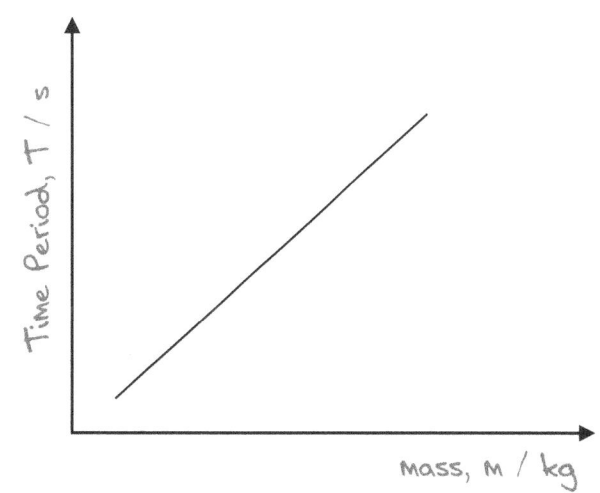

Separate and clearly identify the independent, dependent and controlled variables.

e.g.   Independent variable - mass, m (kg)

       Dependent variable   - time period, T (s)

       Controlled variables - length of spring, l (m)

                         - spring constant, k (Nm⁻¹)

                         - amplitude of oscillation, A (m)

                         - friction (air resistance), F (N)

                         - acceleration due to gravity, g (ms⁻²)

The variables are written in LaTeX form:

- mass, m (kg)
- time period, T (s)
- length of spring, l (m)
- spring constant, k (Nm^{-1})
- amplitude of oscillation, A (m)
- friction (air resistance), F (N)
- acceleration due to gravity, g (ms^{-2})

Aspect 2 – Controlling the Variables:

For the variables identified in Design (Aspect 1 – Selecting Variables), a method needs to be designed which clearly shows how each variable will be controlled. It is important that an experiment is a fair test, so **one** variable at a time is changed (independent variable) and data is collected for a second variable (dependent variable). This will allow a conclusion to be drawn as to whether the change in the independent variable caused any change in the dependent variable. All other variables must remain unchanged so that a fair test can be carried out.

e.g.   <u>Mass, m</u>

The mass on the end of the spring will be changed from 100g to 1 kg, 100g at a time.

<u>Time period, T</u>

The time period of the oscillating spring will be measured using a digital stopwatch. The spring will be timed for 10 oscillations and this time will be divided by 10 to improve accuracy. This procedure will be repeated 3 times and an average taken.

<u>Length of spring, l</u>

As more mass is added, the spring's extension will increase, however, the amplitude will be kept constant

<u>Spring constant</u>

The spring constant will remain constant each time the mass is changed. It will be controlled by using the same spring.

<u>Amplitude of oscillation, A</u>

The amplitude will remain constant each time the mass is changed. It will be controlled using a metre rule.

<u>Friction (air resistance), F</u>

Air resistance is difficult to control. However, it can be minimised by releasing the spring with small amplitude. The ceiling fans will be turned off during the experiment and the same location will be used throughout.

<u>Acceleration due to gravity, g</u>

As long as the experiment is performed at the same location then acceleration due to gravity will remain constant.

Aspect 3 – Developing a method for collecting data:

A list of all the necessary materials is usually needed here. However, if the apparatus is assembled in a particular way then including a diagram may help to understand the assembly.

e.g.   Materials List:

Retort Stand, clamp and boss, Spring, Mass holder (100g), 10 x 100g slotted masses, Metre rule, Stopwatch,

Diagram:

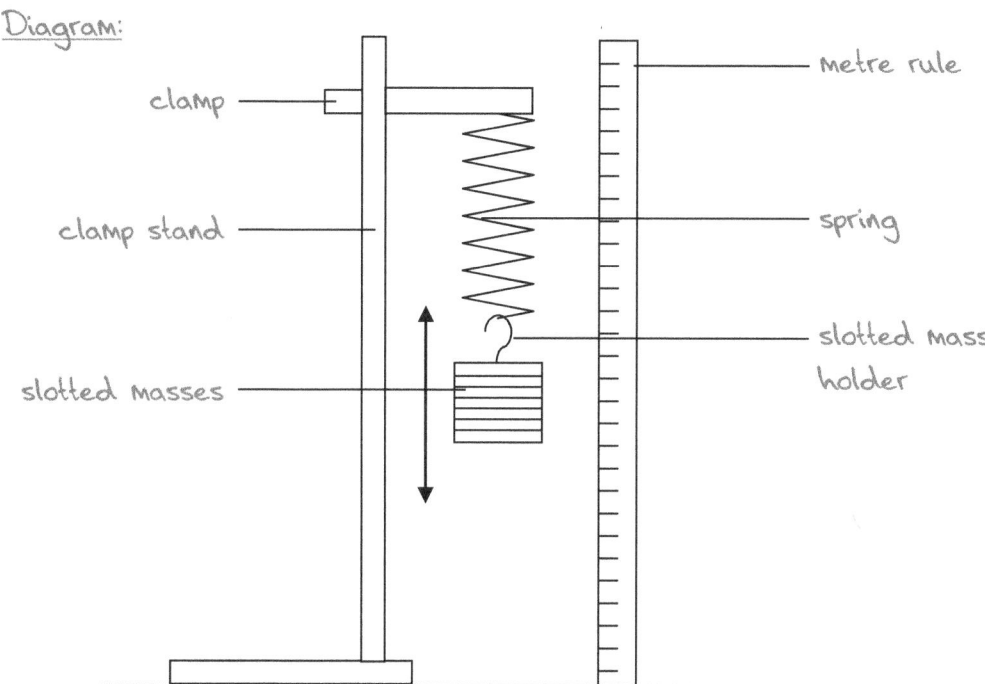

A method should clearly show how you intend to collect data. This should be written in the third person. Bullet points or numbered steps should be used to allow for easy interpretation of your method. Include in your method, any apparatus that you will use and how you will use it. You should also include any techniques that you plan to use in order to reduce random and/or systematic errors.

e.g.
1. Set up the apparatus as shown in the diagram, with one end of the spring attached to the horizontal support on the clamp stand.
2. Attach the slotted mass holder to the bottom of the spring.
3. Position the metre rule next to (but not touching) the spring so that the amplitude of oscillation can be controlled.
4. Pull the 100g slotted mass holder down by 5cm, using a ruler and release the spring.
5. Time 10 complete oscillations of the spring using the stopwatch and record the data.
6. Repeat step 4 two more times (3 times in total).
7. Record the data each time and take an average.
8. Add 100g to the slotted mass holder and repeat steps 4 – 6 until the mass on the slotted mass holder is 1 kg

Data Collection and Processing (DCP)

Aspect 1 – Recording raw data
Records appropriate quantitative and associated qualitative raw data, including units and uncertainties where relevant. The uncertainties in the measuring apparatus used should be accounted for and included in the column headings using the accepted notation e.g. $\Delta m = \pm 0.005$ kg. Significant digits in the recorded raw data should be consistent with the precision of the measuring apparatus used.

Make sure that you present your raw data clearly (usually in the form of a table) with quantities and units at the top of each column.

e.g.   Table 1 - Mass on the spring and time period

Mass, m / kg	Δm / kg	Time for 10 oscillations of the spring / s $\Delta T = \pm 0.01$ s				Time for 1 oscillation / s $\Delta T = \pm 0.01$ s
		T_1 / s	T_2 / s	T_3 / s	Average time T_{AVE} / s	T / s
0.100	0.001	2.30	2.34	2.26	2.30	0.23
0.200	0.002	3.29	3.27	3.25	3.27	0.33
0.300	0.003	4.04	4.09	4.12	4.08	0.41
0.400	0.004	4.65	4.62	4.72	4.66	0.47
0.500	0.005	5.19	5.21	5.13	5.18	0.52
0.600	0.006	5.69	5.66	5.72	5.69	0.57
0.700	0.007	6.10	6.12	6.03	6.08	0.61
0.800	0.008	6.59	6.50	6.53	6.54	0.65
0.900	0.009	6.93	6.88	6.87	6.89	0.69
1.000	0.010	7.31	7.25	7.31	7.29	0.73

Aspect 2 – Processing Raw Data
In IB Physics, data is always considered to be quantitative therefore generally a graph will be constructed which should clearly show a relationship between the dependent and independent variables. When constructing the axes of your graph, include the origin (0,0) so that a fuller understanding of the relationship can be established.

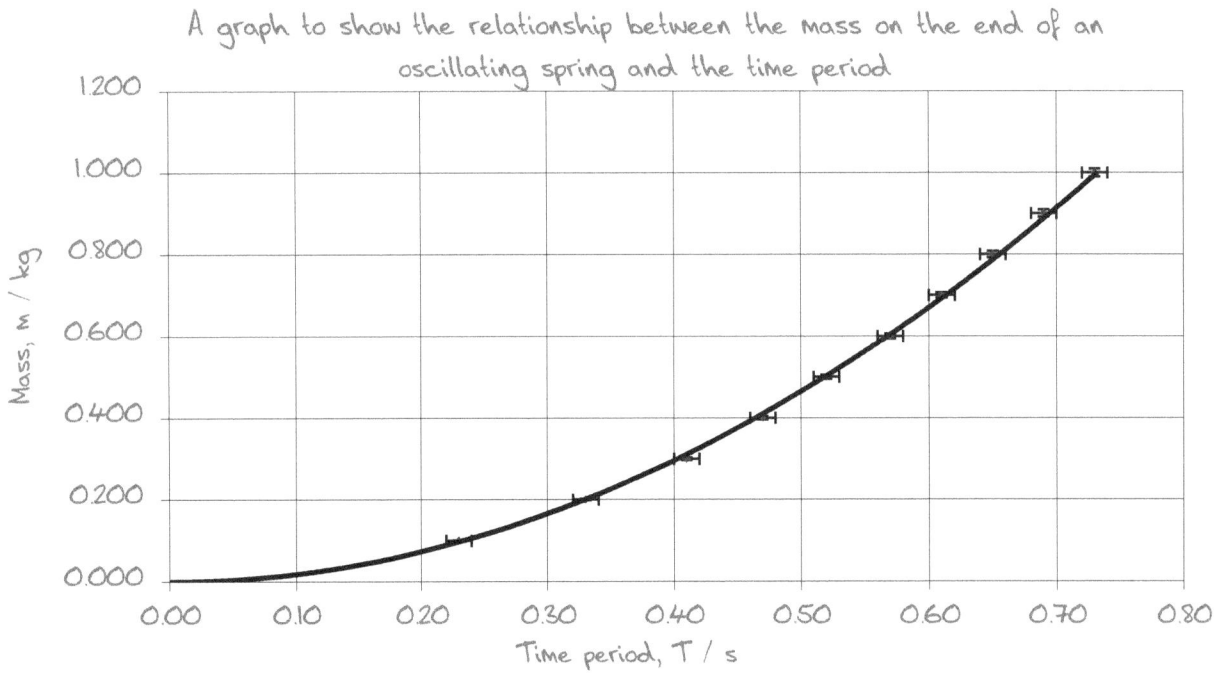

A graph to show the relationship between the mass on the end of an oscillating spring and the time period

Often, more processing will be needed so that a relationship between the two selected variables can be established.

e.g.   The relationship between mass and time period for an oscillating spring appears to be non-linear. The data will therefore be processed in order to find a relationship between these two variables. Since the graph appears to be parabolic, the data will be manipulated so that a graph of mass vs. time^2 can be plotted.

Aspect 3 – Presenting Processed Data

Conventionally, the dependent variable is plotted on the x-axis with the independent variable on the y-axis. It is important that the axes of your graph are clearly labelled and that units are included. A line of best fit should be constructed through the data points so that there is an even distribution of points above and below the line. Uncertainty bars should also be included and higher level students will need to use these uncertainty bars to construct lines of maximum and minimum slope in order to estimate the uncertainty in the slope of the graph.

Data Processing for Higher Level students

e.g.   Table 2 – Mass on the spring and time period squared, T^2.

Mass, m / kg $\Delta m = \pm 0.001$ kg	Time, T / s $\Delta T = \pm 0.01$ s	uncert. in T / %	Time^2, T^2 / s^2	uncert. in T^2 / %	uncert. in T^2 / s^2
0.100	0.23	4.35	0.053	8.70	0.0046
0.200	0.33	3.03	0.109	6.06	0.0066
0.300	0.41	2.44	0.168	4.88	0.0082
0.400	0.47	2.13	0.221	4.26	0.0094
0.500	0.52	1.92	0.270	3.84	0.0104
0.600	0.57	1.75	0.325	3.50	0.0114
0.700	0.61	1.64	0.372	3.28	0.0122
0.800	0.65	1.54	0.423	3.08	0.0130
0.900	0.69	1.45	0.476	2.90	0.0138
1.000	0.73	1.37	0.533	2.74	0.0146

e.g.

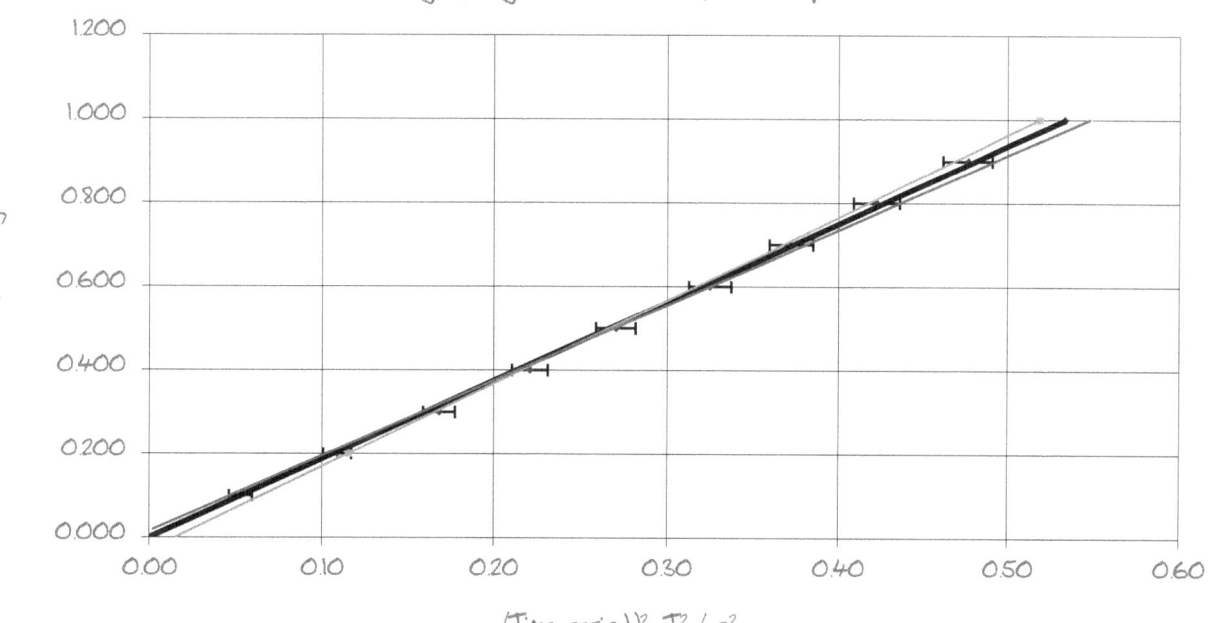

A graph to show the relationship between the mass, m, on the end of an oscillating spring and the time period squared, T^2.

Analysis of the graph:

Gradient of best fit line $\qquad = \dfrac{y_2 - y_1}{x_2 - x_1} = \dfrac{1.00}{0.53} = 1.87 \, kgs^{-2}$

Gradient of steepest line $\qquad = \dfrac{y_2 - y_1}{x_2 - x_1} = \dfrac{1.00}{0.51} = 1.99 \, kgs^{-2}$

Gradient of shallowest line $\qquad = \dfrac{y_2 - y_1}{x_2 - x_1} = \dfrac{0.98}{0.54} = 1.80 \, kgs^{-2}$

From this second graph it can be seen that the mass on the end of an oscillating spring is directly proportional to the time period squared.

Mathematically, $\qquad\qquad T^2 \propto M$

After investigation it was found that the formula relating the time period of an oscillating spring to the mass is

$$T = 2\pi \sqrt{\frac{M}{k}}$$

re-arranging this equation

$$T^2 = \frac{4\pi^2}{k} \cdot M$$

which is of the form $\qquad\qquad y = Mx \qquad$ (equation of a straight line)

The gradient of the best fit line is therefore equal to $4\pi^2/k$, the spring constant can now be found:

$$gradient = 1.87 = \frac{k}{4\pi^2}$$
$$k = 1.87 \times 4\pi^2$$
$$k = 73.8 \, kgs^{-2} \, (Nm^{-1})$$

The range of uncertainty in this value can be calculated using the maximum and minimum lines on the graph

$$max \; gradient = 1.99 = \frac{k}{4\pi^2} \qquad\qquad min \; gradient = 1.80 = \frac{k}{4\pi^2}$$
$$k = 1.99 \times 4\pi^2 \qquad\qquad\qquad\qquad k = 1.80 \times 4\pi^2$$
$$k = 78.6 \, kgs^{-2} \, (Nm^{-1}) \qquad\qquad\quad k = 71.1 \, kgs^{-2} \, (Nm^{-1})$$

therefore the spring constant, $k = 73.8 \, ^{+4.8}_{-2.7} \, Nm^{-1}$

Conclusion and Evaluation

Aspect 1 – Concluding
Start the conclusion by restating the aim or research question. This can immediately lead into whether you have successfully accomplished the aim and, from this, whether your hypothesis was correct. Always refer to the data to back-up any statements. Explain the results using scientific reasoning and (where possible) try to compare the results of your experiment with published results and/or values.

e.g.   The aim of this experiment was to investigate how the mass on the end of a spring affects the time period of oscillation. I predicted that the time period would be directly proportional to the mass of the oscillating system. This prediction turned out to be incorrect as the graph of mass vs. time period is clearly non-linear. The second graph of mass vs. time period squared however turned out to be linear and therefore I can conclude that mass is directly proportional to the time period squared.

After further investigation this conclusion is supported as the equation for the time period of an oscillating spring is

$$T = 2\pi\sqrt{\frac{M}{k}}$$

re-arranging this equation

$$T^2 = \frac{4\pi^2}{k} \cdot M$$

so

$$T^2 \propto M$$

The gradient of the straight line was then used to calculate the spring constant, k for the spring used in this experiment.

This was found to be k = $73.8 \, ^{+4.8}_{-2.7}$ Nm^{-1}

This cannot be compared to a literature value, but a simple Hooke's Law experiment can be used to verify the validity of this result. When the spring was loaded to 10N, the extension was seen to be 14.2 cm (0.142 m). From this the spring constant can be determined.

$$F = kx$$

$$k = \frac{F}{x} = \frac{10}{0.142} = 70.4 \text{ Nm}^{-1}$$

Comparing the two results, a percentage deviation can be calculated.

$$\text{Percentage deviation} = \frac{73.8 - 70.4}{73.8} = 4.6\%$$

This is an acceptable difference (within 5%) and therefore the experiment can be assumed to be successful and the results accurate.

Aspect 2 – Evaluating Procedure(s)

The evaluation of procedure(s) should take account of any limitations and weaknesses in the original plan when the data was collected and should include an evaluation of the apparatus used to collect the data. This section also requires comments to be made regarding errors (random and systematic) in the results.

e.g.   In general the method and apparatus worked well. There were however some modifications that were made when collecting the data that were not stated in the original plan.

1.   Parallax error (random error) when reading the ruler was accounted for by placing the recorders eye level with the bottom of the spring.

2.   When lighter loads were used (100g and 200g), the period of oscillation was so fast that it was quite difficult to count ten oscillations. In these cases, more than three trials were recorded and the then three were selected to use in the table of results based on how close they were to each other. This was a random error that caused uncertainty in the period.

3.   The slotted masses were never checked for their accuracy. The 100g stamped on each was taken to be accurate. This may have produced a systematic error, depending on how inaccurate the masses were and the consistency of their inaccuracy

4.   I found that the 1kg mass was a very large mass to use. It was so heavy that it caused the whole system, to bounce around uncontrollably. This was a random error that caused uncertainty in the period.

Aspect 3 – Improving the investigation

Suggests realistic improvements in respect of identified weaknesses and limitations.

e.g.   The investigation could have yielded more accurate results if the following modifications were made in future;

1.   Parallax error. A horizontal pin could have been stuck to the bottom of the slotted mass holder. This would have pointed at the ruler so that the elimination of parallax error simply by guess would have been eliminated.

2.   Timing issues relating to human reaction could have been avoided if the "Vernier" ultra-sonic motion detector was placed below the oscillating spring. The data would have therefore been collected automatically and human interaction would have been avoided.

3.   A balance should have been used to check to accuracy of each of the 100g slotted masses.

4.   For next time, the range of masses should be changed to 50g - 500g in increments of 50g.

INVESTIGATIONS

INVESTIGATING MURPHY'S LAW

Aim:
Murphy's Law states that if there are a number of possible outcomes to a situation, the actual outcome will be the worst possible one! One application of this law is breakfast. If you drop your toast, it will always land jam side down.

Apparatus:
Anything you need to perform the experiment to your plan.

Design:
Design a procedure to test Murphy's Law that includes appropriate use of apparatus for the control, collection and analysis of data.

This procedure should include the following sections.

- Defining the Problem and selecting variables:
- Controlling the Variables:
- Developing a method for collecting data:

Include a quantitative hypothesis for your investigation.

IB Criteria Assessed
Design,
Data Collection and Processing,
Conclusion and Evaluation

Criteria assessed	Aspect			Level awarded
	1	2	3	
D				
DCP				
CE				

Data Collection and Processing:
- Record the raw data (both quantitative and qualitative) for the experiment in a suitable form. Include uncertainties due to the precision of the measuring apparatus.
- Process your quantitative raw data.
- Present the processed data in an appropriate way and include errors and uncertainties

Conclusion and Evaluation:
- Draw conclusions based on your interpretation of the data
- Evaluate your own plan, including any weaknesses and/or limitations. Include an evaluation of the apparatus used.
- In light of the weaknesses and limitations suggested above, suggest ways in which the procedure could be modified in order to improve it for the future.

INVESTIGATING THE FALL OF A COFFEE FILTER

Aim:
As a basket-type coffee filter falls, it tends to fall straight down and not flip over. This allows us to design a lab with several interesting variations on the independent and dependent variables. You very recently were told to drop coffee filters and change several different variables as you dropped the filters. You also measured several different responding variables. Based on your brief exposure to the wonders of falling coffee filters, you will design a lab that could be performed by another student in a normal physics class.

Apparatus:
Basket-type coffee filters and any other equipment you think you might need.

IB Criteria Assessed
Design,
Data Collection and Processing,
Conclusion and Evaluation

Criteria assessed	Aspect			Level awarded
	1	2	3	
D				
DCP				
CE				

Design:
Design a procedure to test a factor(s) that affects a falling coffee filter and that includes appropriate use of apparatus for the control, collection and analysis of data

This procedure should include the following sections.

- Defining the Problem and selecting variables:
- Controlling the Variables:
- Developing a method for collecting data:

Step by step instructions and diagrams are helpful to the reader and highly recommended. Also include a hypothesis and a sketch graph of what you think will happen.

Data Collection and Processing:
- Record the raw data (both quantitative and qualitative) for the experiment in a suitable form. Include uncertainties due to the precision of the measuring apparatus.
- Process your quantitative raw data.
- Present the processed data in an appropriate way and include errors and uncertainties

Conclusion and Evaluation:
- Draw conclusions based on your interpretation of the data
- Evaluate your own plan, including any weaknesses and/or limitations. Include an evaluation of the apparatus used.
- In light of the weaknesses and limitations suggested above, suggest ways in which the procedure could be modified in order to improve it for the future.

INVESTIGATING THE SIMPLE PENDULUM

Aim:

This investigation is designed to give you an introduction to IB design, data collection, data processing and evaluation experiments.

You should design an experiment in order to test different variables that may contribute to the time period of a pendulum and try to determine a mathematical relationship between these variables.

Apparatus:

Whatever you feel is suitable for this experiment.

IB Criteria Assessed

Design,
Data Collection and Processing,
Conclusion and Evaluation

| Criteria | Aspect | | | Level |
assessed	1	2	3	awarded
D				
DCP				
CE				

Design:

Design a procedure to test a factor(s) that affects the time period of a Simple Pendulum that includes appropriate use of apparatus for the control, collection and analysis of data. This procedure should include the following sections.
* Defining the Problem and selecting variables:
* Controlling the Variables:
* Developing a method for collecting data:

Include a quantitative hypothesis for your investigation.

Data Collection and Processing:

* Record the raw data (both quantitative and qualitative) for the experiment in a suitable form. Include uncertainties due to the precision of the measuring apparatus.
* Process your quantitative raw data.
* Present the processed data in an appropriate way and include errors and uncertainties

Conclusion and Evaluation:

* Draw conclusions based on your interpretation of the data
* Evaluate your own plan, including any weaknesses and/or limitations. Include an evaluation of the apparatus used.
* In light of the weaknesses and limitations suggested above, suggest ways in which the procedure could be modified in order to improve it for the future.

INVESTIGATING UNCERTAINTIES – MEASURING INSTRUMENT CIRCUS

Aim:
You need practice in taking all measurements with a ± error and then performing calculations to arrive at a final answer in the form, for example:

Volume = 28.9 cm^3 ± 2 cm^3

or   Volume = 28.9 cm^3 ± 6.9 %

There are also some useful "tricks" you can learn to improve the accuracy of your experiments. Think about these methods and include them when collecting your data.

IB Criteria Assessed
Data Collection and Processing,

Criteria assessed	Aspect 1	2	3	Level awarded
D				
DCP				
CE				

Apparatus:
Metre rule, Vernier callipers, micrometer, marble, bag of rice, digital scales, mechanical balance, short length of copper wire, 20 sheets of paper, stop watch, ping pong ball, measuring cylinder, tea cup.

Method:
You must use the available instruments to measure the following items:

1. the diameter of the marble.
2. the surface area of the marble.
3. the volume of the marble.
4. the mass of a grain of rice.
5. the diameter of the copper wire.
6. the time for a ping pong ball to fall from a height of 3 metres.
7. the volume of liquid in a tea cup.
8. the area of the rectangle below.

Data Collection and Processing:
- Record the raw data (both quantitative and qualitative) for the experiment in a suitable form. Include uncertainties due to the precision of the measuring apparatus with a ± heading for each column on your table.
- Remember that your ± error estimates are not the only sources of error in the experiments. Look for and think about other causes of errors - e.g.: is the wire the same thickness along the whole length?. There are several problems like this in the experiments. If you spot them, your experimental results will be more accurate.
- What assumptions did you make in your results? e.g. - are all rice grains the same size?
- Process your quantitative raw data.
- Specify what instrument you used and the methods you employed to improve the accuracy of your results.

INVESTIGATING ERRORS AND UNCERTAINTIES IN EXPERIMENTS

Aim:
All experiments are done with measuring instruments. No instrument is perfectly accurate - they all have limits to their accuracy. It is important that you realise that no experiments give perfect, exact answers. To illustrate this you are going to measure the density of a slide with:

1. a precision electronic balance, and a Vernier gauge.
2. then with a lever arm balance and a metre rule.

IB Criteria Assessed
Data Collection and Processing,

Criteria	Aspect			Level
assessed	1	2	3	awarded
D				
DCP				
CE				

From these 2 sets of measurements you will calculate the density of the slide.
One of the experiments is more accurate than the other, but both are imprecise. From now on, all your experimental results will include the uncertainties associated with the measuring apparatus used. (and for HL student, a calculation of $\pm$ error in the readings and in the final result).

Apparatus:
Vernier callipers, metre rule, microscope slide, electronic scale, lever arm balance.

Method:
1. Measure the mass of the slide on both the precise and less precise scale.
2. Measure the length, breadth and thickness of the slide with the vernier and then the metre rule.
3. Write all of your measurements in a suitable table of results with a $\pm$ uncertainty at the top of each column.

Theory:
$$density = \frac{mass}{volume}$$

Calculate density using the readings from the metre rule and less precise balance. Repeat this with the readings from the precise scale and the vernier callipers. Compare the precision of these two results.

Data Collection and Processing:
Calculate the maximum and the minimum possible value for the density of the slide based upon the precision of the two sets of apparatus.

Using the uncertainty in the readings, calculate the largest and smallest possible values for the density of the slide. In both cases produce an answer for the density of the slide in the form:

Density = ***** g cm^{-3} $\pm$ ** g cm^{-3}
Density = ***** g cm^{-3} $\pm$ ** %

All measuring instruments have limits to their accuracies but you can make them less accurate by not using them carefully. It is important that :

1. the balances are at zero before the slide is put on them.
2. you **put** the slide on the balance; don't **drop** it on them.
3. the metre rule is not worn away at one end.

In all your future experiments you should always be aware of the importance of using your measuring instruments as accurately as possible.

DETERMINING STIFFNESS OF STEEL BY THE OSCILLATIONS OF A HACKSAW BLADE

Aim:

This is to show that a concept difficult to measure can be easily calculated indirectly from a suitable equation. Also this is further practice in plotting an appropriate graph and using the gradient to find a constant - in this case the constant is the stiffness (E) of the steel. Also you will see how to use your graph to obtain a $\pm$ estimate of accuracy.

Apparatus:

Clamp, hacksaw blade, stop watch, 2 magnets, ruler, micrometer, digital scales, 2 blocks of wood.

IB Criteria Assessed

Data Collection and Processing, Conclusion and Evaluation

| Criteria | Aspect | | | Level |
assessed	1	2	3	awarded
D				
DCP				
CE				

Diagram:

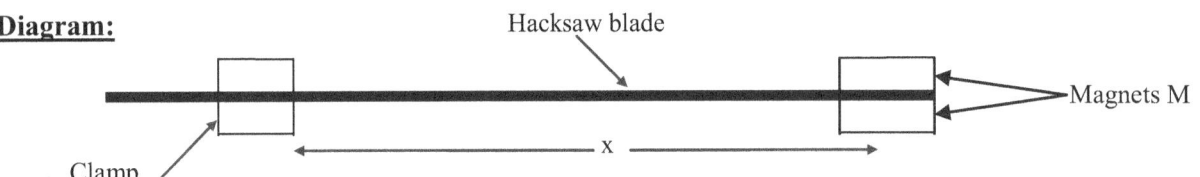

Hacksaw blade

Magnets M

Clamp

x

Method:

Clamp the hacksaw blade between the 2 blocks of wood with the blade <u>vertical</u>. Attach the 2 magnets to the <u>end</u> of the blade, about 0.15 m from the blocks (distance x). Pull the magnets to one side and release them so that the blade vibrates. Time the period T. Repeat this for various lengths of x up to 0.3 m. Also measure the mass of the magnets (M), the thickness of the hacksaw blade (d), and the breadth (b).

Data Collection and Processing:

Period T, mass M and stiffness E are related by the equation:

$$T^2 = \left[\frac{16\pi^2 M}{bd^3 E}\right] x^3$$

- Draw a suitable table to include all of the variables together with their actual uncertainties and percentage uncertainties due to the apparatus used.
- Plot a suitable graph that will allow you to find the stiffness E.
- What is the accuracy of your result? To estimate this, find the best-fit line on the graph and the worst fit lines.
- Calculate a value for E from each gradient and so find a value for the uncertainty in your result for E.

Conclusion and Evaluation:

- What does the term "stiffness" mean?
- What is the value of the stiffness (E) for steel?
- How does this value compare with the one that you have obtained.
- Suggest reasons for any differences.
- Suggest areas where the procedures used in this practical may have been the cause of some of these errors.
- Suggest modifications to the practical to minimise any errors and shortcomings.

INVESTIGATING THE STOPPING DISTANCE OF A BICYCLE

Aim:
Safety on the road is an important part of everyday life. Cars are capable of faster and faster speeds and these higher speeds bring the need for better brakes, better tyres, and better roads.

In an emergency situation, perhaps a person stepping off the curb and into the road, a driver has to be able to stop the vehicle before reaching the person. There are numerous factors which might affect this stopping distance.

Your task is to identify these factors, choose ONE to test and find a relationship between your chosen variable and the stopping distance.

IB Criteria Assessed
Design,
Data Collection and Processing,
Conclusion and Evaluation

Criteria assessed	Aspect			Level awarded
	1	2	3	
D				
DCP				
CE				

Apparatus:
One bicycle plus anything else you need to perform the experiment to your plan.

Design:
Design a procedure to test a factor that affects the stopping distance of a bicycle that includes appropriate use of apparatus for the control, collection and analysis of data.

This procedure should include the following sections.

- Defining the Problem and selecting variables:
- Controlling the Variables:
- Developing a method for collecting data:

Include a quantitative hypothesis for your investigation.

Data Collection and Processing:
- Record the raw data (both quantitative and qualitative) for the experiment in a suitable form. Include uncertainties due to the precision of the measuring apparatus.
- Process your quantitative raw data.
- Present the processed data in an appropriate way and include errors and uncertainties

Conclusion and Evaluation:
- Draw conclusions based on your interpretation of the data. Include a comparison with published data relating the stopping distance of a vehicle and the variable that you tested.
- Evaluate your own plan, including any weaknesses and/or limitations. Include an evaluation of the apparatus used.
- In light of the weaknesses and limitations suggested above, suggest ways in which the procedure could be modified in order to improve it for the future.

INVESTIGATING THE TORSIONAL PENDULUM

Aim:
The time period of a pendulum was first investigated by Christian Huygens in the 17th Century (after apparently observing a swinging chandelier in church). From Huygen's work the first accurate clocks were invented and refined

You may be very familiar with traditional swinging pendulums and the variables that do and do not affect their time periods. In this design lab, you will learn about the workings of a torsional (twisting) pendulum.

IB Criteria Assessed
Design,
Data Collection and Processing,
Conclusion and Evaluation

Criteria assessed	Aspect			Level awarded
	1	2	3	
D				
DCP				
CE				

Diagram:

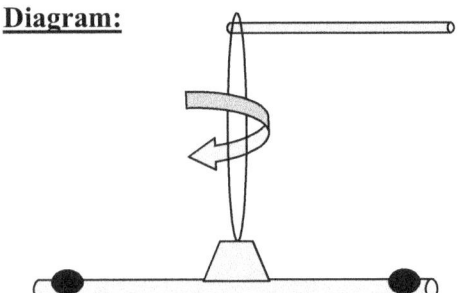

Apparatus:

- A torsional pendulum made from a 30cm acrylic ruler and elastic band (shown opposite)
- Anything else you might find in a normal physics classroom.

Design:
Design a procedure to test how a certain variable (of your choice) may affect the rate of an oscillating torsional pendulum that includes appropriate use of apparatus for the control, collection and analysis of data. As always, this should design lab should include:

- Defining the Problem and selecting variables:
- Controlling the Variables:
- Developing a method for collecting data:

Step by step instructions and diagrams are helpful to the reader and highly recommended.

Also include a quantitative hypothesis for your investigation together with a sketch graph of what you think will happen.

Data Collection and Processing:
- Record the raw data (both quantitative and qualitative) for the experiment in a suitable form. Include uncertainties due to the precision of the measuring apparatus.
- Process your quantitative raw data.
- Present the processed data in an appropriate way and include errors and uncertainties

Conclusion and Evaluation:
- Draw conclusions based on your interpretation of the data. Include a comparison of your results with any published data regarding torsional pendulums. Compare your results with your original hypothesis.
- Evaluate your own plan, including any weaknesses and/or limitations. Include an evaluation of the apparatus used.
- In light of the weaknesses and limitations suggested above, suggest ways in which the procedure could be modified in order to improve it for the future.

INVESTIGATING PROJECTILES

Aim:
To calculate the initial horizontal take off velocity of an object that is allowed to fall after being released from a ramp.

IB Criteria Assessed
Data Collection and Processing,
Conclusion and Evaluation

Criteria	Aspect			Level
assessed	1	2	3	awarded
D				
DCP				
CE				

Method:
1. Set up the apparatus as shown below.
2. Place the wooden board in the vertical position so that it is touching the bottom of the ramp.
3. Release the ball bearing from height h so that it accelerates down the ramp and makes a mark on carbon paper attached to the wooden board.
4. Move the wooden board away from the ramp by a small distance (s_x) and release the ball once again.
5. Repeat for 6 to 8 distances of s_x.

Diagram:

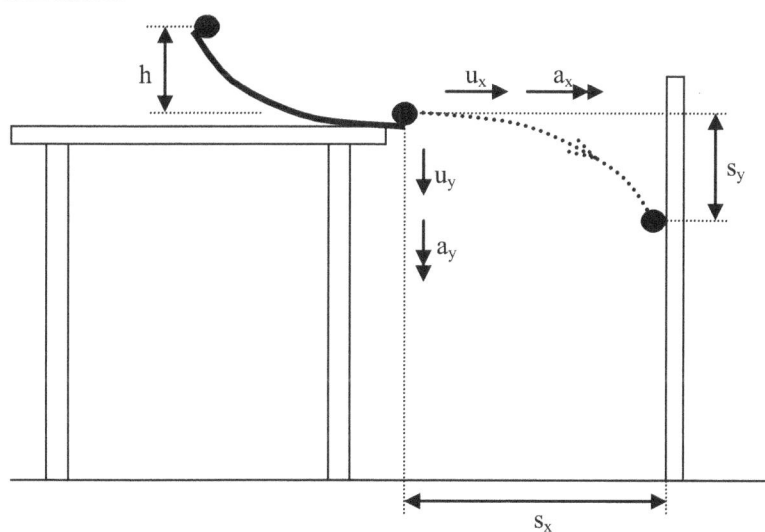

h = vertical height for the ball to be dropped.

u_x = initial horizontal velocity.

u_y = initial vertical velocity.

a_x = initial horizontal acceleration.

a_y = initial vertical acceleration.

s_x = horizontal distance.

s_y = vertical distance.

Data Collection and Processing:
*   Measure distances s_x and corresponding distances s_y and record these values in a suitable table. **Note: It is important to use fundamental units.**
*   Include uncertainties due to the precision of the measuring apparatus.
*   Process your quantitative raw data by drawing a suitable graph which will allow the initial horizontal take off velocity u_x to be calculated.
*   Present the processed data in an appropriate way and include errors and uncertainties

Conclusion and Evaluation:
*   Evaluate the data that you have collected and analysed - compare your results by calculating the initial horizontal take off velocity (u_x) using the conservation of mechanical energy equations ($mgh = \frac{1}{2}mv^2$)
*   Evaluate the procedure, including any modifications you had to make to overcome problems. Include an evaluation of the apparatus used.
*   Suggest ways in which the procedure could be modified in order to improve it for the future.

INVESTIGATING FORCES IN EQUILIBRIUM

Aim:
1. To use vector addition to calculate the mass of an unknown object.
2. To gain an understanding of balanced forces in situations of static equilibrium when two or more forces act at a point in a system.

Equipment:
2 pulleys mounted on a board, thread, slotted masses, 3 sheets of white paper for each student.

Method:
1. Set up the apparatus as shown in the diagram
2. Hang the unknown mass and two slotted masses from the 3 threads and let the system come to equilibrium.
3. Displace it several times and notice the variation in the position of the point O.
4. When the system comes to rest in an approximately mean position, accurately mark the position of point O and the directions of the 3 forces on a piece of white paper fixed to the board.
5. Label the point of intersection of the 3 force vectors as O.
6. Repeat the above procedure for two more systems by changing the magnitude of coplanar forces acting on the body.
7. Record the value of forces A and B for each system.

Data Collection and Processing:
1. Attach your three A4 sheets which clearly show the original data obtained.
2. Using a suitable technique, determine the vector sum A+B and hence calculate the mass of the unknown object.
3. Compare this sum with the value of C obtained by measuring the unknown mass with a balance.
4. Include any uncertainties (both magnitude and direction).

IB Criteria Assessed
Data Collection and Processing

Criteria assessed	Aspect			Level awarded
	1	2	3	
D				
DCP				
CE				

Diagram:

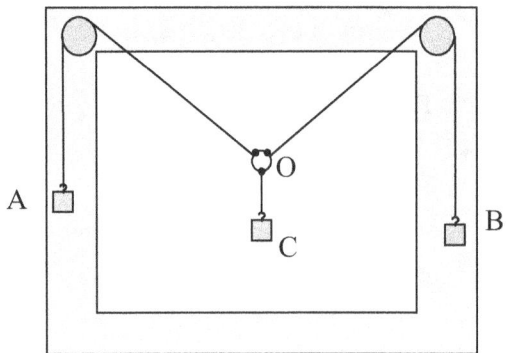

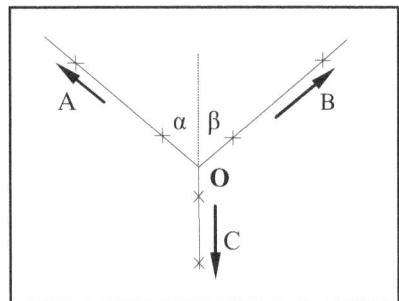

INVESTIGATING ERRORS AND UNCERTAINTIES USING ACCELERATION DUE TO GRAVITY

Aim:
This experiment is designed to introduce you to the detail that is needed when constructing and presenting your table of results for an experiment. It will also allow you to practice some of the data analysis techniques that have been recently discussed in class.

IB Criteria Assessed
Data Collection and Processing, Conclusion and Evaluation

Criteria	Aspect			Level
assessed	1	2	3	awarded
D				
DCP				
CE				

Method:
1. For various heights, drop the object as it falls to the floor, accelerating at a rate of $g = 10 m/s^2$.
2. Time the object 3 times as it falls and take an average of these times.

Theory:
The acceleration due to gravity for an object, falling close to the Earth's surface is given by

$$T = 2\pi\sqrt{\frac{h}{g}}$$

Data Collection and Processing:
- Collect and record pairs of data (height and time) including units and uncertainties.
- Present these data clearly in a suitable table.
- Process your raw data in a way which will allow you to accurately calculate the value of g (acceleration due to gravity) – HINT Use the theory above to manipulate your data to give a linear relationship
- Take into account any errors or uncertainties in your processed data.
- Draw a suitable graph that will allow an analysis to be made on the raw data.
- From your graph calculate the acceleration due to gravity, g, for a falling object.
- Include actual and percentage uncertainties by plotting suitable straight lines to determine these errors.

Conclusion and Evaluation:
- Make valid conclusions related to the value calculated for g found. Compare this calculated value of g to literature values.
- Evaluate the method, including any modifications you had to make to overcome problems
- Include an evaluation of the apparatus used.
- Suggest ways in which the procedure (and apparatus) could be modified in order to improve future investigations.

INVESTIGATING THE FLIGHT OF AN ELASTIC BAND

Aim:
A very important idea in scientific experiments is "controlling the variables". The aim is to keep all factors constant in the experiment except for the 2 you are testing. In this practical you will learn, for example, to examine the relationship between angle of launch and range while at the same time keeping other factors, such as amount of stretch of the band, constant. This is called a "fair test" in science.

Also, because of the randomness of the flight of the elastic band, you will also realise the importance of repeating your readings.

IB Criteria Assessed
Design,
Data Collection and Processing,
Conclusion and Evaluation

Criteria assessed	Aspect 1	2	3	Level awarded
D				
DCP				
CE				

Apparatus:
An elastic band plus any other equipment that you think you need.

Design:
Design a procedure to investigate the flight of your elastic band that includes appropriate use of apparatus for the control, collection and analysis of data. This procedure should include the following sections.
- Defining the Problem and selecting variables:
- Controlling the Variables:
- Developing a method for collecting data:

Include a quantitative hypothesis for your investigation.

Data Collection and Processing:
- Record the raw data (both quantitative and qualitative) for the experiment in a suitable form. Include uncertainties due to the precision of the measuring apparatus.
- Process your quantitative raw data.
- Present the processed data in an appropriate way and include errors and uncertainties

Conclusion and Evaluation:
- Draw conclusions based on your interpretation of the data
- Evaluate your own plan, including any weaknesses and/or limitations. Include an evaluation of the apparatus used.
- In light of the weaknesses and limitations suggested above, suggest ways in which the procedure could be modified in order to improve it for the future.

INVESTIGATING WORK DONE AND ENERGY TRANSFERRED ON AN INCLINED PLANE

Aim:
1. To find a relationship between force applied and distance moved up an inclined plane
2. To use this relationship to prove that **Work done = Energy transferred**
3. To use raw data as an exercise in data manipulation in order to calculate the height, h

IB Criteria Assessed
Data Collection and Processing, Conclusion and Evaluation

Criteria assessed	Aspect			Level awarded
	1	2	3	
D				
DCP				
CE				

Diagram:

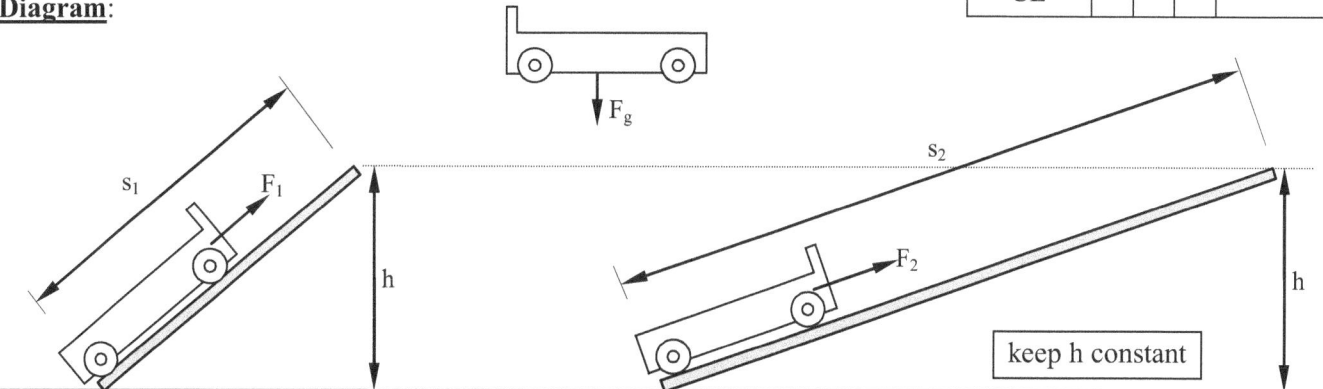

Method:
1. Set up the apparatus as shown in the diagram below (any value for h will do).
2. Measure the height of the ramp, h and record it.
3. Measure the weight of the dynamics trolley, F_g and record it.
4. Using a force meter, pull the trolley up the ramp (maintaining a constant velocity), recording the length of the ramp, s, and the force required, F.
5. Change the angle of the ramp and hence the distance, s and force, F. **KEEP h CONSTANT.**
6. Repeat steps 2 to 5 above until you are satisfied with the quantity of collected data.

Data Collection and Processing:
- Collect and record pairs of results for s and F including units and uncertainties.
- Present these data clearly.
- Process your raw data in a way which will allow you to accurately calculate the value of h (the height of the ramp). Refer to the "Aim" of the investigation.
- Include any errors or uncertainties in your processed data.

Conclusion and Evaluation:
- Give a conclusion and explanation of your results. Compare your calculated value of h to the actual value (measured from the apparatus)
- Evaluate the above procedure (method) and apparatus used, including limitations, weaknesses or errors.
- Suggest ways of improving the investigation.

INVESTIGATING THE BALLISTIC PENDULUM

Aim:
To investigate the laws of conservation of momentum and energy.

Theory:
When a projectile is fired at the pendulum and captured by it, the combination of pendulum and projectile will swing upwards by an amount which is indirectly related to the original velocity of the bullet. According to the law of conservation of energy, the increase in height, Δh, of the pendulum block and projectile is such that the increase in potential energy is equal to the kinetic energy just after the collision.

IB Criteria Assessed
Data Collection and Processing, Conclusion and Evaluation

Criteria assessed	Aspect			Level awarded
	1	2	3	
D				
DCP				
CE				

Diagram:

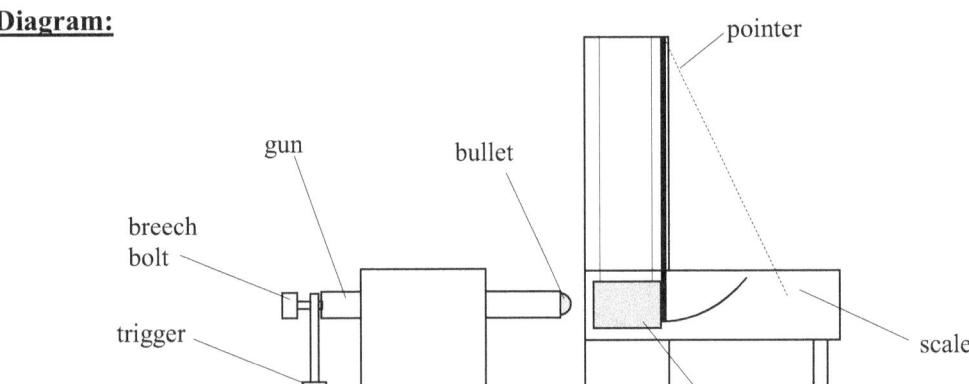

Apparatus:
Ballistic pendulum (model P62401)

Method:
1. Measure the mass of the pendulum and the mass of the bullet using an electronic balance.
2. Measure the length of the pendulum with a ruler.
3. Pull the breech bolt until it is latched by the trigger at the 1st position, load the gun with the bullet and fire the gun so that the bullet gets lodged inside the pendulum
4. Record the position of the pointer in degrees.
5. Repeat step 3, 5 times using the same notch on the breech bolt
6. Repeat steps 3, 4 and 5, using the 2nd and 3rd latch positions

Data Collection and Processing:
- In a suitable table record all of the data collected.
- Include actual and percentage uncertainties due to the measuring apparatus used.
- Analyse the raw data in an appropriate way that will accurately determine the velocities of the bullet at the 3 breech bolt positions.

Conclusion and Evaluation:
- Evaluate the results of the experiment. Suggest ways of checking the reliability of your analysis.
- Evaluate the procedure, including any modifications you had to make to overcome problems. Include an evaluation of the apparatus used.
- Suggest ways in which the procedure could be modified in order to improve it for the future.

                                        Original Lab Sheet by Mike Dickinson

INVESTIGATING ENERGY TRANSFER AND ENERGY LOSS OF A ROLLING BALL

Aim:

In many situations in physics very useful results can be obtained by considering energy changes. Also in these changes, some energy is always "lost" and it is important to be aware of this and to attempt to explain what has happened to it. In this case you are going to examine a change from potential energy to kinetic energy when a ball rolls down a slope and from this find a value for "g" the acceleration due to gravity.

Also you will learn how to manipulate 2 equations so that a useful graph can be plotted from the resulting equation.

IB Criteria Assessed
Data Collection and Processing

Criteria assessed	Aspect			Level awarded
	1	2	3	
D				
DCP				
CE				

Apparatus:

Ball bearing, stop clock, metre rule, ramp.

Diagram:

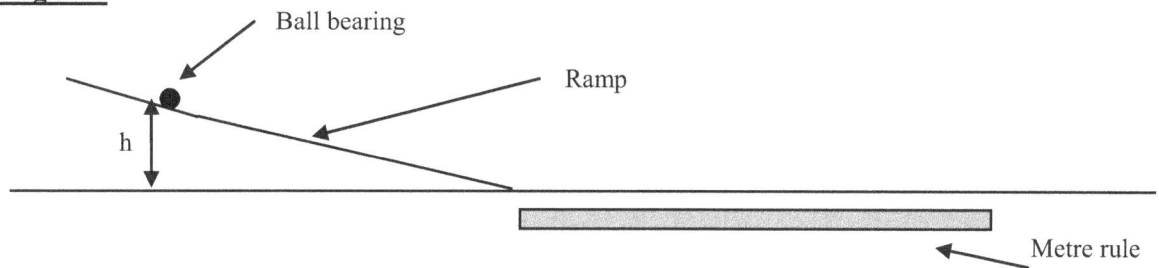

Method:

1. Release the ball bearing from a height "h"
2. Record the time taken for it to travel 1m once it leaves the ramp (or 2 m or 3 m if you think that this will be better).
3. Repeat this for several different heights.

Theory:

The equation for potential energy is $P.E. = mgh$

The equation for kinetic energy is $K.E. = \frac{1}{2}mv^2$

Data Collection and Processing:

- Collect and record pairs of results for h and t including units and uncertainties.
- Present these data clearly.
- Combine the 2 equations above and from them plot a suitable graph from which you can calculate a value for "g".
- Include any errors or uncertainties in your processed data.

Try to explain where energy has been "lost" and why this leads to a different value for "g".

INVESTIGATING NEWTON'S 2ND LAW OF MOTION USING A TICKER TIMER

Aim:
To use the ticker timer and ticker tape to verify Newton's 2nd Law of Motion (Law of Inertia)

IB Criteria Assessed
Data Collection and Processing, Conclusion and Evaluation

Criteria assessed	Aspect			Level awarded
	1	2	3	
D				
DCP				
CE				

Apparatus:
1 runway, 9 x 10g masses plus holder, string, scissors, dynamics trolley, tickertape and timer, lab power supply, 2 x long wires, wedge, pulley, top pan balance.

Method:
1. Set up the apparatus as shown below. Connect the ticker timer to a 12V a.c. supply.
2. The slope of the runway should be adjusted so that the trolley runs at a constant speed down the slope when pushed (as judged by the eye). In this condition there is no resultant force acting on the trolley along the slope. The runway has now been "friction compensated"
3. Measure the mass of the trolley
4. Cut a length of tickertape (about 1m), attach it to the trolley.
5. Hang a mass holder, as shown. This has a mass of 10g and therefore pulls the trolley with a force of 0.098N down the slope. With the ticker timer running, allow the trolley to be pulled down the slope. CATCH THE TROLLEY BEFORE IT FALLS OFF THE END OF THE RUNWAY.
6. Repeat the above using different masses to create different pulling forces.

Diagram:

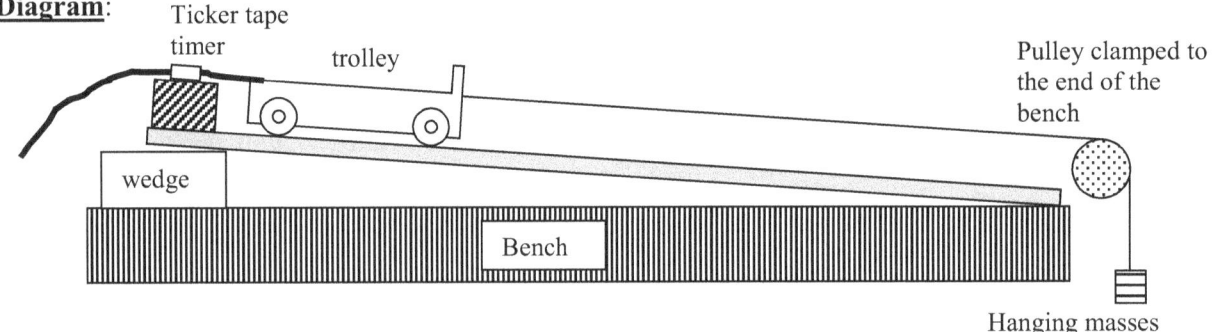

Ticker tape timer
trolley
Pulley clamped to the end of the bench
wedge
Bench
Hanging masses

Theory:
Distance $s = ut + \frac{1}{2}at^2$ $\therefore$ $a = 2s / t^2$ (initial velocity $u = 0$)
Ticker timer produces 50 "ticks" per second $\therefore$ time between ticks = 0.02s (50 ticks = 1s)
Measuring a length of tape which is x ticks long will allow you to calculate the actual acceleration.

Data Collection and Processing:
- Collect data from the ticker tape and record the forces which were used for each piece of tape. Present these data clearly.
- Calculate the expected acceleration in each case using Newton's 2nd Law.
- Use the ticker timers to determine the actual accelerations.

Conclusion and Evaluation:
- Give a conclusion and explanation of your results, compare the two values for acceleration in each case and make appropriate comments.
- Evaluate the above procedure (method) and apparatus used, including limitations, weaknesses or errors.
- Suggest ways of improving the investigation for the future.

INVESTIGATING HOOKE'S LAW

Robert Hooke was one of the first to notice a relationship between the force applied to an elastic object and its extension.

Aim:
This lab is designed to test and verify Hooke's Law which states that "the extension of an elastic material is directly proportional to the applied force so long as the elastic limit is not exceeded".

Apparatus:
Clamp and stand, spring, meter rule, mass holder, slotted masses

Method:
1. Set up the apparatus as shown in the diagram.
2. Apply various forces to the spring
3. Record its extension
4. (Remember that the extension = final length − original length).

Data Collection and Processing:
- Record pairs of data for force and extension in a suitable table.
- Your results table and the presentation of data should include any uncertainties associated with the apparatus that you have used.
- Use your results to plot a suitable graph which will allow you to accurately calculate the spring constant, k, for your spring.
- SL & HL, include uncertainty bars for each data point.
- HL, calculate the uncertainty in your final value for the spring constant by drawing maximum and minimum slope on your graph.

IB Criteria Assessed
Data Collection and Processing, Conclusion and Evaluation

Criteria assessed	Aspect			Level awarded
	1	2	3	
D				
DCP				
CE				

Diagram:

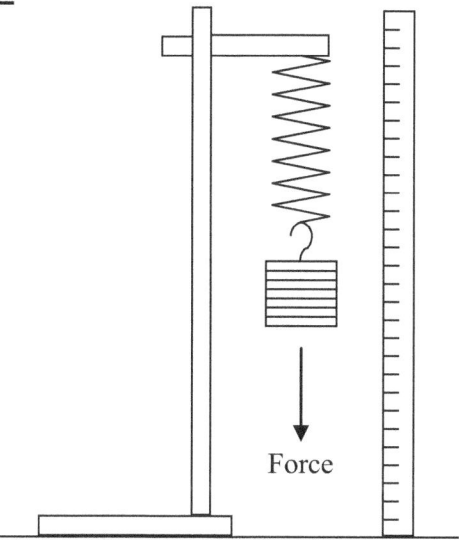

Force

Conclusion and Evaluation:
- Your evaluation should include a final value for the spring constant of your spring together with an estimate of the uncertainty.
- Does your graph verify Hooke's Law?
- Evaluate the procedure and result including limitations, weaknesses or errors.
- Suggest possible causes of errors and modifications that could be introduced to improve the investigation.

INVESTIGATING SPRINGS IN SERIES AND PARALLEL

Aim:
To investigate the interaction of various spring systems. (Examples of different spring systems are shown in the diagrams below).

IB Criteria Assessed
Design,
Data Collection and Processing,
Conclusion and Evaluation

Diagrams (Explanation of series and parallel spring systems):

Criteria assessed	Aspect			Level awarded
	1	2	3	
D				
DCP				
CE				

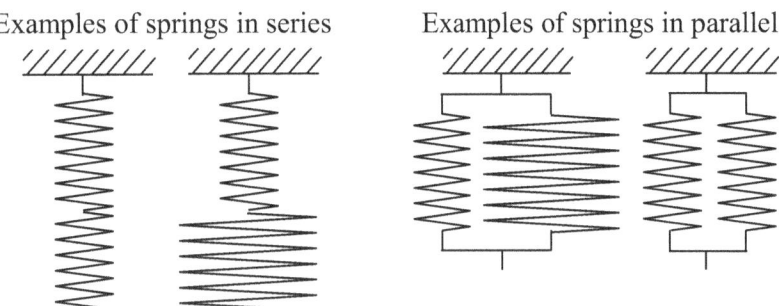

Examples of springs in series   Examples of springs in parallel

Apparatus:
Various springs of different spring constants. Any other apparatus that you can think of to complete the investigation.

Design:
Design a procedure that includes appropriate use of apparatus for the control, collection and analysis of data. This procedure should include the following sections

Data Collection and Processing:
- Show the results for the experiment in a suitable table. Include uncertainties.
- Present all of your results in the form of suitable graphs
- Suggest relationships to help explain the systems analysed.

Include a quantitative hypothesis for your investigation.

Conclusion and Evaluation:
- Evaluate the data that you have collected and analysed - compare your results to literature values.
- Evaluate your own plan, including any modifications you had to make to overcome problems. Include an evaluation of the apparatus used.
- Suggest ways in which the procedure could be modified in order to improve it for the future.

     Original Lab Sheet by Mike Dickinson

INVESTIGATING CIRCULAR MOTION

Aim:
1. To investigate circular motion.
2. To use the formula for centripetal acceleration to calculate the mass of a rotating object.

Apparatus:
Grass tube, rubber cork, string, meter rule, stopclock, masses and mass holder.

Theory:
When the system is in equilibrium $F_R = F_C$
(speed and radius of rotation remain constant)

Method:
1. Set the force, F_R, hanging from the string and record it.
2. Rotate the string and record the radius, r, and time period, T.
3. Change the force, F_R and repeat until 6 to 8 sets of data are recorded.

IB Criteria Assessed
Data Collection and Processing, Conclusion and Evaluation

Criteria	Aspect			Level
assessed	1	2	3	awarded
D				
DCP				
CE				

Diagram:

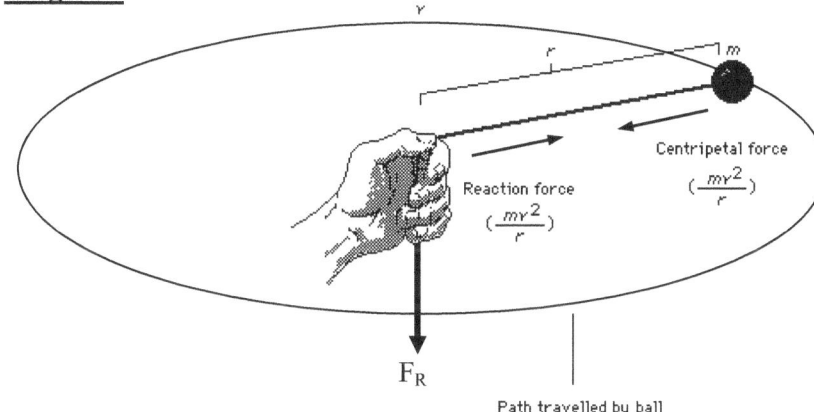

F_R = Reaction Force
m = mass of ball
v = speed
r = radius of rotation
T = time for 1 rotation

Data Collection and Processing:
- In a suitable table record all of the data collected.
- Include actual and percentage uncertainties due to the measuring apparatus used.
- Draw a suitable graph that will allow you to use the formula for centripetal acceleration in order to calculate the mass of the rotating object. Include your error bars to indicate the uncertainty in the apparatus used. HL include a maximum and minimum gradient in order to determine the overall uncertainty in your calculated answer

Conclusion and Evaluation:
- Evaluate the results of the experiment - compare your results by taking the mass of the object using a balance.
- Evaluate the procedure, including any modifications you had to make to overcome problems. Include an evaluation of the apparatus used.
- Suggest ways in which the procedure could be modified in order to improve it for the future.

INVESTIGATING PING - PONG BALLS

Aim:

This is a real - life problem. At Wimbledon tennis tournament, the balls are kept in a fridge before the players use them. This is because the ball's performance changes enough with temperature to affect their bounciness and this could have a serious effect on the result of the game - enough maybe for example to just put the ball over the line.

You will learn that investigating a real - life situation can become very complicated. "How hard do table tennis players hit the ball?" It is often very difficult to simulate in the laboratory what actually happens in the real world.

IB Criteria Assessed

Design,
Data Collection and Processing,
Conclusion and Evaluation

Criteria	Aspect			Level
assessed	1	2	3	awarded
D				
DCP				
CE				

Apparatus:

3 ping - pong balls plus any other equipment you think will be necessary.

Design:

This is an open ended experiment so you can do anything you want, but remember that you will have to show to the "International Ping - Pong Federation" that you have thoroughly investigated all aspects of the problem and that you are certain of your conclusions. Design a procedure that includes appropriate use of apparatus for the control, collection and analysis of data. This procedure should include the following sections:

- Defining the Problem and selecting variables:
- Controlling the Variables:
- Developing a method for collecting data:

Include a quantitative hypothesis for your investigation.

Data Collection and Processing:

- Show the results for the experiment in a suitable table. Include uncertainties.
- Analyse your raw data in such a way that will allow a conclusion to be reached about the relationship between your selected variable and the bounciness of the Ping-Pong balls.

Conclusion and Evaluation:

- Evaluate the data that you have collected and analysed - compare your results to literature values if possible.
- Evaluate your own plan, including any modifications you had to make to overcome problems. Include an evaluation of the apparatus used.
- Suggest ways in which the procedure could be modified in order to improve it for the future.

DETERMINING SPECIFIC HEAT CAPACITY BY THE ELECTRICAL METHOD

Aim:
1. To calculate the SHC of aluminium, copper, iron and water.
2. To see how accurately you can measure a known quantity.
3. To appreciate the effect of heat loss - this is important in all heat experiments.

IB Criteria Assessed
Data Collection and Processing, Conclusion and Evaluation

Criteria assessed	Aspect			Level awarded
	1	2	3	
D				
DCP				
CE				

Method:
1. Setup the circuit below.
2. Take the initial temperature of the material being tested.
3. Turn on the power supply and set it to 12V. Start the stop watch.
4. Record the temperature, voltage, current at regular time intervals and record them.
5. Continue to record data for 20 minutes.
6. Repeat the above method for the other two materials under test

Diagram:

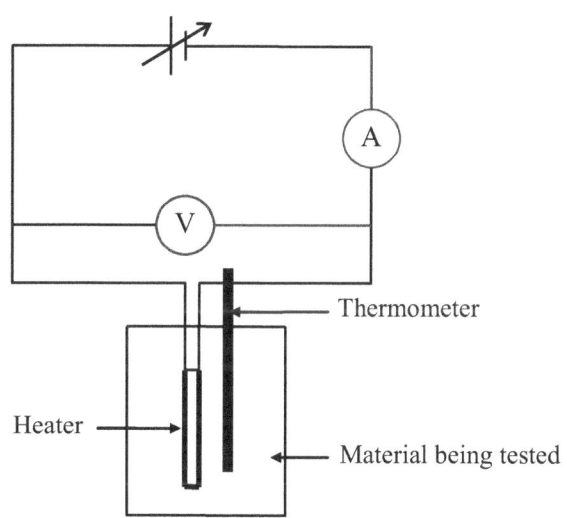

Apparatus:
Aluminium block,
copper block,
iron block,
beaker of water,
power supply,
thermometer,
ammeter,
voltmeter,
connecting leads,
stop watch,
heater,
digital balance.

Theory:

Energy supplied by heater	=	Energy received by aluminium block
V I t	=	**m c ΔT**

Data Collection and Processing:
- Record temperature, voltage, current and time in a suitable table.
- Your results table and the presentation of data should include any uncertainties associated with the apparatus that you have used.
- Draw a suitable graph that will allow you to use the formula above to calculate the specific heat capacities of the three materials.

Conclusion and Evaluation:
- Your conclusion should include values for the specific heat capacities of the 3 different materials and a comparison with literature values.
- Evaluate the method including any modifications you had to make to overcome problems. Include an evaluation of the apparatus used.
- Suggest ways in which the procedure could be modified in order to improve it for the future.

DETERMINING SPECIFIC HEAT CAPACITY BY THE METHOD OF MIXTURES

Aim:
1. To calculate the SHC of aluminium, copper, iron and water
2. To see how accurately you can measure a known quantity.
3. To appreciate the effect of heat loss - which is important in all heat experiments.

IB Criteria Assessed
Data Collection and Processing, Conclusion and Evaluation

Criteria assessed	Aspect			Level awarded
	1	2	3	
D				
DCP				
CE				

Method:
1. Measure the mass of the material under test
2. Heat the water and test material in a beaker and keep boiling for about 5 minutes (to ensure the test material is at 100°C)
3. Measure the initial temperature and mass of the cold water in the other beaker
4. Move the test material from the hot water to the cold water
5. Stir the water
6. Wait until the water, test material and beaker reach thermal equilibrium
7. Take the final temperature of the mixture

Diagram:

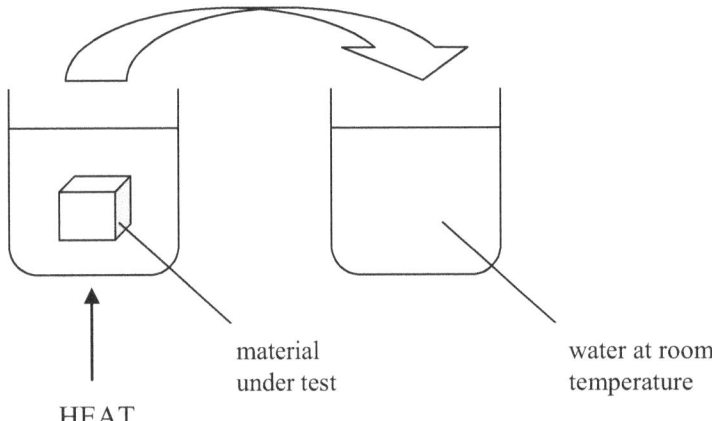

material under test

HEAT

water at room temperature

Apparatus:
aluminium block
copper block
iron block
2 beakers of water
thermometer
Bunsen burner / heater
balance

Theory:

> **Energy supplied by test material   =   Energy received by water and beaker**

Data Collection and Processing:
- Record your results in a suitable table.
- Your results table and the presentation of data should include any uncertainties associated with the apparatus that you have used.
- Calculate the specific heat capacity of the three test materials.

Conclusion and Evaluation:
- Your conclusion should include values for the specific heat capacities of the 3 different materials and a comparison with literature values.
- Evaluate the method including any modifications you had to make to overcome problems. Include an evaluation of the apparatus used.
- Suggest ways in which the procedure could be modified in order to improve it for the future.

INVESTIGATING SPECIFIC LATENT HEAT OF VAPORISATION OF WATER

Aim:
This experiment is to show you that we can do scientific experiments with non - specialised equipment (in this case an ordinary kettle) and still obtain reasonable results. You will be able to compare your result with the accepted literature value. Also, this experiment is useful as one that can be done in a very short time to obtain an answer that is fairly close to the true value - which is a useful skill in science.

IB Criteria Assessed
Data Collection and Processing, Conclusion and Evaluation

Criteria assessed	Aspect			Level awarded
	1	2	3	
D				
DCP				
CE				

Apparatus:
Electric kettle, scales, stop watch.

Diagram:

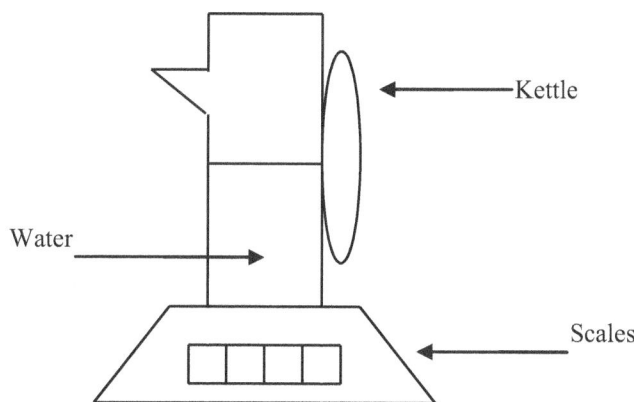

Method:
Put the kettle, ½ full of water on the scales. Switch it on and wait for the water to start to boil. When it is clearly boiling, start the stop watch and record the mass of the kettle every 30 seconds. Continue to take mass measurements until the mass of the kettle has fallen by 50 g. Repeat the above procedure a number of times to minimise uncertainties.

Theory:

> **Energy supplied by the kettle = Energy received by the water**

From the above equation you can calculate the specific latent heat of vaporisation of water (the power of the kettle is written on it).

Data Collection and Processing:
- Record all your data in a suitable table of results.
- Record the uncertainties in your measurements
- Using the equation given, plot a suitable graph that will allow you to calculate the Specific Latent Heat of Vaporisation of Water.
- Include an estimate of the uncertainty in your calculated value

Conclusion and Evaluation:
- Give a conclusion and explanation of your results; compare to literature values.
- Evaluate the above procedure (method) and apparatus used, including limitations, weaknesses or errors.
- Identify any weaknesses and suggest ways of improving the investigation.

INVESTIGATING SPECIFIC LATENT HEAT OF FUSION OF ICE

Aim:
All heat experiments have problems with heat loss or gain from the surroundings. This experiment contains a trick to try and get round this difficulty. The water is pre - heated 5 ^{0}C above room temperature and then cooled to 5 ^{0}C below room temperature. In this way, during the first half of the experiment the water is losing heat and in the second half it is gaining the same quantity of heat - the loss and gain of heat cancel each other out. Tricks like this are very important in experiments in order to overcome practical difficulties.

IB Criteria Assessed
Data Collection and Processing, Conclusion and Evaluation

Criteria assessed	Aspect			Level awarded
	1	2	3	
D				
DCP				
CE				

Method:
Measure room temperature. Weigh the empty beaker. Add water to the beaker and weigh them again. Warm the water to 5 ^{0}C above room temperature. Add small pieces of crushed ice, stirring all the time, until the temperature falls to 5 ^{0}C below room temperature. Weigh the beaker and ice to determine the mass of ice added. Repeat the experiment to get 2 sets of results.

Diagram:

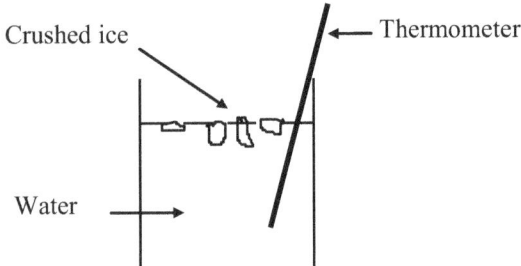

Apparatus:
Beaker
thermometer
crushed ice
scales
Bunsen burner or kettle.

Theory:

Energy received by ice = Energy released by water

From this equation you can calculate the specific latent heat of fusion of ice.

Data Collection and Processing:
- Record temperature, mass of water and mass of ice in a suitable table.
- Your results table and the presentation of data should include any uncertainties associated with the apparatus that you have used.
- Process your date in such a way that the Specific Latent Heat of Fusion for the ice can be established

Conclusion and Evaluation:
- Your evaluation should include values for the specific latent heat of fusion for ice and a comparison with literature values.
- Were your 2 results very different?
- You should also comment on the methods and apparatus used, and suggest possible causes of errors and modifications that could be introduced to improve the investigation.

INVESTIGATING THE POWER AND TEMPERATURE OF THE SUN

Aim:
In many situations it is impossible to measure something directly. This experiment is a clever example of how to use physics theory, and some mathematics, to measure indirectly, what is impossible to do directly - namely to find the power and temperature of the sun.

Method:
1. Measure the mass of the test tube and the mass of the water.
2. Take the temperature of the water.
3. Focus the sun on to the water and stir occasionally.
4. Hold the lens and test tube steady until the water has increased in temperature a reasonable amount (do not allow the water to boil).

IB Criteria Assessed
Data Collection and Processing, Conclusion and Evaluation

Criteria assessed	Aspect			Level awarded
	1	2	3	
D				
DCP				
CE				

Diagram:

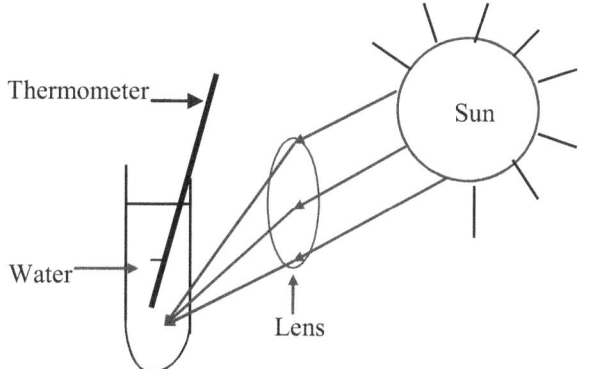

Thermometer

Sun

Water

Lens

Apparatus:
Magnifying glass
Thermometer
Test tube
Stopclock.

Theory:
The energy that passes through the lens heats the water and the test tube. This energy can be calculated using the equation for specific heat capacity. This is the energy passing through area A m^2 of the lens. Knowing the radius of orbit of the Earth around the Sun, you can calculate the total energy emitted by the Sun. The surface temperature of the Sun can be calculated by using the "Stefan - Boltzmann" equation.

> Energy / second / m^2 from the Sun's surface = $5.7 \times 10^{-8} \times$ (Kelvin temperature)^4

Data Collection and Processing:
- Collect data that will allow you to calculate the surface temperature and power of the sun.
- Include units and uncertainties.
- Present these data clearly.
- Using the above method and theory, calculate the surface temperature and power of the sun.

Conclusion and Evaluation:
- The information you need can all be found in your data book (the true value for the power of the Sun is 3.9×10^{26} W). Compare your calculated value to this literature value
- There are many faults in this experiment. Try to think of some of them. If you have time then repeat the experiment but with an improved method.
- Evaluate the method including any modifications you had to make to overcome problems. Include an evaluation of the apparatus used.
- Suggest ways in which the procedure could be modified in order to improve it for the future.

INVESTIGATING CHARLES' LAW AND ABSOLUTE ZERO

Aim:

Physics has many hundreds of laws and these can be tested in experiments to see if they are true. In 1787 the French scientist, J.A.C. Charles, published a law connecting the volume and temperature of gases. Your task is to see if you agree with his law. Another thing you have to consider is "is my experiment accurate enough to prove or disprove the law?" It could be that the law is good but your experiment isn't. This process of testing a proposed law with an experiment is one of the foundations of all science.

IB Criteria Assessed

Conclusion and Evaluation

Criteria assessed	Aspect			Level awarded
	1	2	3	
D				
DCP				
CE				

Apparatus:

Beaker, thermometer, capillary tube with plug of concentrated sulphuric acid, 30 cm ruler, Bunsen burner.

Diagram:

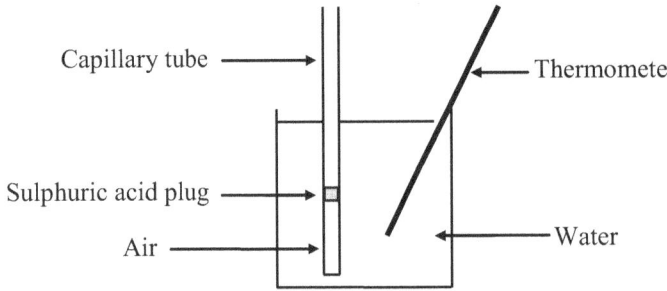

Method:

Put the capillary tube in the beaker as shown. Measure the length of the air below the plug in the capillary with a ruler. Slowly warm the water and approximately every 10 ^{0}C, measure the length of air again. Repeat this until the water is boiling. Make sure that the air column in the capillary tube is always below the level of the water.

Data Collection and Processing:

- Collect pairs of data for length of air and temperature
- Plot a graph of "length of air" against "temperature in °C".
- Extrapolate the line until it reaches the x axis (y=0)

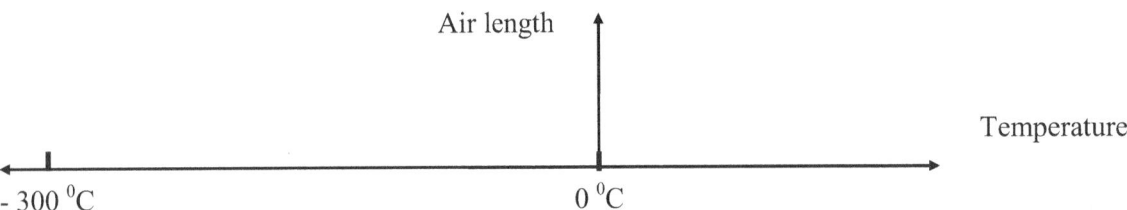

Conclusion and Evaluation:

- Give a conclusion and explanation of your results; compare your results to literature values.
- Evaluate the above procedure (method) and apparatus used, including limitations, weaknesses or errors.
- Identify any weaknesses and suggest ways of improving the investigation.

INVESTIGATING RATE OF COOLING

Is the rate of cooling in still air $\propto$ (excess temperature)$^{5/4}$?

Aim:

You have a law expressed mathematically and there are 2 useful skills here for you to learn:
1. You will need to use your knowledge of logs to manipulate the data so that a linear relationship can be established.
2. Having collected your data you need to plot the appropriate graph in order to test the 5/4 relationship.

Apparatus:

Beaker of water, heater, stopclock, thermometer.

Method:

1. Heat the water to boiling point
2. Measure and record the temperature of the water
3. Start the stopclock and remove the water from the heat
4. Record the temperature of the water each minute as it cools

Hint:

Notice that the law says "rate of cooling". This means "°C per minute". This is important when you decide what graph you are going to plot to test the law of cooling. Remember also that before you decide if the law of cooling is true or not, you must decide if your experiment was accurate enough to justify your conclusions.

Data Collection and Processing:

*   Collect pairs of date for temperature and time as the water cools
*   Show the results for the experiment in a suitable table. Include uncertainties.
*   Analyse your raw date in such a way that will allow a conclusion to be reached about the relationship between rate of cooling and excess temperature.

Conclusion and Evaluation:

*   Evaluate the data that you have collected and analysed - compare your results to the value given.
*   Evaluate your own plan, including any modifications you had to make to overcome problems. Include an evaluation the apparatus used.
*   Suggest ways in which the procedure could be modified in order to improve it for the future.

IB Criteria Assessed
Data Collection and Processing,
Conclusion and Evaluation

Criteria assessed	Aspect			Level awarded
	1	2	3	
D				
DCP				
CE				

INVESTIGATING THE PRESSURE / VOLUME RELATIONSHIP FOR A BALLOON

Aim:

This experiment is different because you have to try and formulate some sort of relationship between the pressure and the volume of a balloon but you have no idea what it might be so you will have to confront a completely original problem for which there is no answer in a text book. Good luck!

Method:

1. Using first the round balloon, partially inflate it and connect it to the manometer.
2. The difference in height of water between the 2 sides gives the pressure (use "$p = \rho \times g \times h$" to convert to Pascals).
3. How to measure the volume of the balloon is your decision.
4. You need to use a bit of ingenuity here.
5. Inflate the balloon a bit more and once again measure pressure and volume.
6. Obtain a total of 5 or 6 readings.
7. Now do the whole experiment again but this time using the "sausage" balloon.

IB Criteria Assessed
Data Collection and Processing

Criteria assessed	Aspect			Level awarded
	1	2	3	
D				
DCP				
CE				

Apparatus:

Manometer, 1 "round" balloon, 1 "sausage" balloon, plus whatever else you think you need.

Diagram:

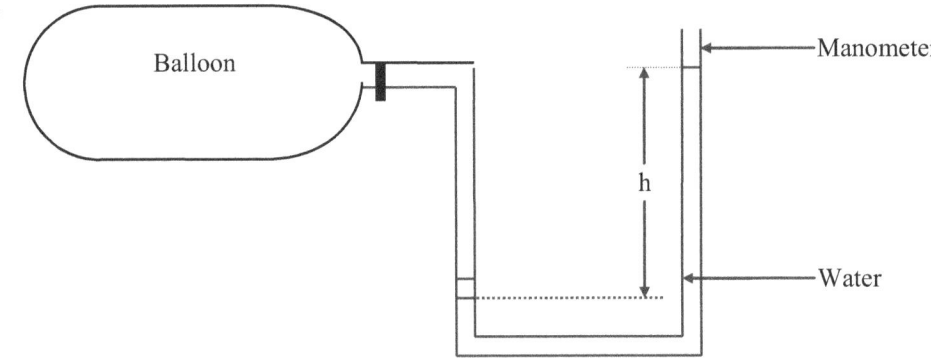

Data Collection and Processing:

*   Collect whatever data you feel is necessary in order to complete the data processing parts below.
*   Plot a suitable graph which will allow you to find the relation between pressure and volume - e.g. "p" against "$v^{1/2}$".
*   Remember that you are looking for a straight line graph.

Do both balloons give the same result?

DETERMINING THE TEMPERATURE OF A WIRE BY EXPANSIVITY

Aim:
To measure the temperature of a hot wire is very difficult to do directly - you can't simply touch a thermometer against the wire. This experiment is interesting because you use one property (thermal expansion) in order to measure another (temperature). This indirect technique is used a lot in science - for example the temperature of a star can be measured by examining its colour.

Also in the real world, engineers have to know the effect of current through wires in order to allow for sag in high voltage overhead cables

IB Criteria Assessed
Data Collection and Processing

Criteria assessed	Aspect			Level awarded
	1	2	3	
D				
DCP				
CE				

Apparatus:
1 m of copper wire, 2 clamp stands, four 100 g masses, variable power supply, ammeter, sheet of plane paper, leads, crocodile clips.

Diagram:

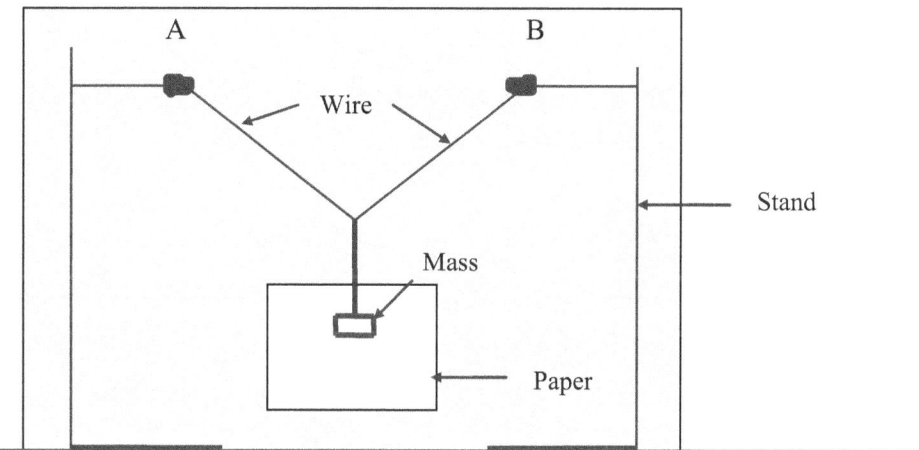

Method:
1. Clamp the wire firmly at A and B using the 100 g masses as grips.
2. Hang the mass from the wire and mark its position on a piece of blank paper.
3. Connect the power supply with the ammeter in series, to A and B.
4. Set the current to 1A and mark the new position of the hanging mass.
5. Repeat this in steps of 1 A, up to 5A.
6. Use the lines on the paper to measure the increase in length of the wire for each reading and, knowing the thermal expansivity of the wire's material (look this up in a data book), calculate the average temperature of the wire for each of the 5 different currents.

Data Collection and Processing:
- Collect and record pairs of data for current and average temperature.
- Present these data in a results table – include the uncertainties in these two variables
- Plot a suitable graph to see the relationship between current and average temperature. Include uncertainties.
- Look up the melting point of copper and use your graph to estimate what current will melt the wire.
- Include an estimate in the uncertainty of your calculated value.

DETERMINING THE REFRACTIVE INDEX OF GLASS BY REAL AND APPARENT DEPTH

Aim:
When you look down at your feet in a swimming pool, they appear to be closer than normal. This is a short experiment that uses this phenomenon to produce a reasonably accurate result for the refractive index of glass. So you have taken an everyday observation and invented an experiment to obtain a <u>quantitative</u> measurement of what you saw and calculate the uncertainty in your final answer.

IB Criteria Assessed
Data Collection and Processing

Criteria assessed	Aspect			Level awarded
	1	2	3	
D				
DCP				
CE				

Apparatus:
Travelling microscope, 2 glass blocks, lycopodium powder.

Diagram:

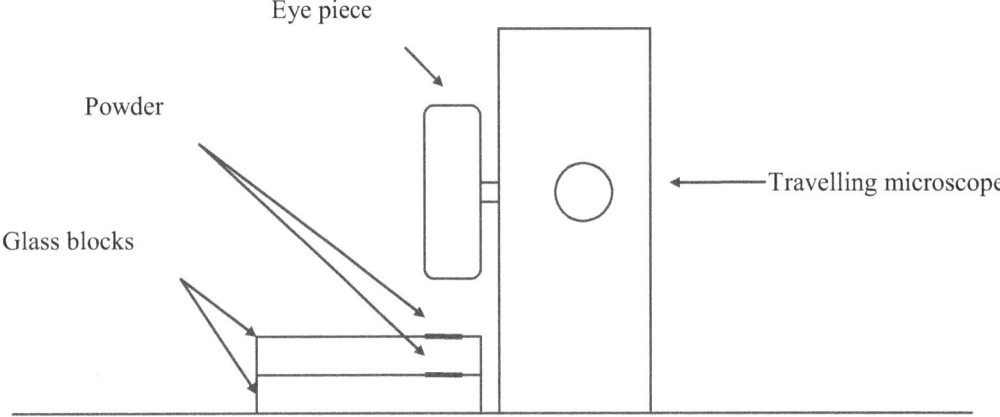

Method:
Put a small amount of powder on the top of the first block and focus the microscope on it - take reading d_1. Place the second block on top of the first and focus on the powder again - take reading d_2. Put a small amount of powder on top of the second block and focus the microscope on it - take reading d_3.

Data Collection and Processing:
- Record the data, as specified in the above method, in a suitable form
- Include units and uncertainties.
- Use your raw data to accurately calculate a value for the refractive index of the glass.
- Include the uncertainty in your final value and show your calculations.

          Original Lab Sheet by Brian Seve – Modified to the current IB syllabus requirements by Mike Dickinson

INVESTIGATING REFRACTION OF LIGHT

Aim:
1. To investigate the relationship between the angles of incidence and refraction as light travels into a rectangular Perspex block.
2. To verify experimentally the formula: Refractive index $n = \sin i / \sin r$
3. To investigate total internal reflection and the "critical angle", C, of a refractive substance.
4. To verify experimentally the formula:
5. Refractive index $n = 1 / \sin C$

IB Criteria Assessed
Data Collection and Processing

Criteria assessed	Aspect			Level awarded
	1	2	3	
D				
DCP				
CE				

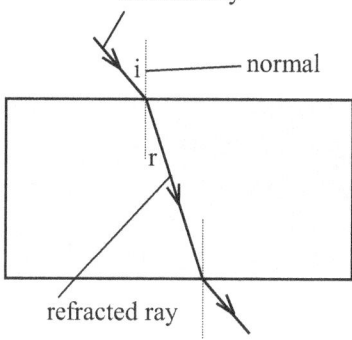

incident ray

normal

refracted ray

r = angle of incidence
i = angle of incidence

Method (Part A):
1. Place the rectangular Perspex block on a sheet of white paper.
2. Draw around the block (in case it gets disturbed).
3. Shine a single, thin ray of light from a ray box, incident to one long face of the block.
4. Mark the incident ray entering the block and the emergent ray exiting the block.
5. Measure the angle of incidence (i) and the corresponding angle of refraction (r).
6. Vary the angle of incidence so that 8 different pairs of results can be collected.

Method (Part B):
1. Place the semi-circular Perspex block on a sheet of white paper.
2. Draw around the block (in case it gets disturbed).
3. Shine a single, thin ray of light from a ray box, incident to the curved face of the block, pointing at the centre of the straight face
4. Mark the incident ray entering the block and the emergent ray exiting the block.
5. Increase the angle of incidence (i) until the emergent ray disappears and the light is reflected inside of the glass block.
6. Measure angle i and the corresponding angle of reflection.

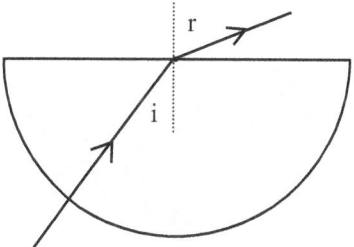

Data Collection and Processing:
- From Part A, collect and record pairs of data (angles i and r) including units and uncertainties.
- Present these data clearly in a suitable table.
- From Part B, record the critical angle C.
- Use your results from part A of the experiment to plot a suitable graph that will verify the formula for refractive index of Perspex.
- Calculate the refractive index of the Perspex using your graph.
- Use the results from part B to calculate the refractive index of the Perspex.
- Compare the results for refractive index from the two experiments.
- Which one do you think is the most accurate? Why?

INVESTIGATING MALUS' LAW

As unpolarized light passes through a polarizing filter, the Polaroid removes all the light whose axes of polarization are not in-line with the plane of polarization of the Polaroid. This leaves only light with electric field oscillations in one orientation. The light is said to be "plane polarized". The intensity of the polarized light is half the original source intensity, $I = \frac{1}{2}I_0$

If a second Polaroid, known as an analyser, is placed in front of the first Polaroid, the intensity of the light can be controlled according to Malus' Law. The intensity is seen to be a function of the angles of polarization between the first and second Polaroids.

IB Criteria Assessed
Data Collection and Processing, Conclusion and Evaluation

Criteria assessed	Aspect			Level awarded
	1	2	3	
D				
DCP				
CE				

Aim:
1. To verify the use of a Polaroid to halve the intensity of the original light source.
2. To verify Malus' Law using crossed Polaroids.

Theory:
The relationship between the intensity of the light leaving the second (crossed) Polaroid is given by Malus' Law.

$$I = I_0 \cos^2 \theta$$

where I = intensity of the transmitted beam
I_0 = intensity of the incident (plane polarized) beam
θ = angle between the plane of polarization of the incident beam and the plane of Polaroid's axis

Diagram:

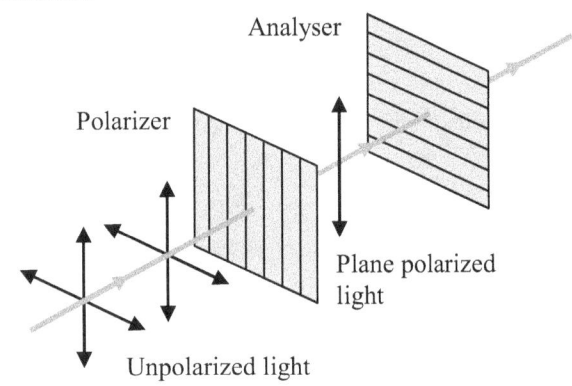

Analyser

Polarizer

Plane polarized light

Unpolarized light

Method:
1. Set up the apparatus as shown in the diagram.
2. Start with the 2 Polaroids giving the same plane of polarization (this can be achieved by lining the 0° mark on both Polaroids at the 12 o'clock position in their holders.
3. Adjust the angle between the Polaroids in 5° increments until their planes of polarization are perpendicular (as shown in the diagram)
4. Record pairs of data for angle and light intensity, until an angle is reached where the intensity of the light falls to a zero (or a minimum).
5. Continue past this minimum – the intensity of the light should start to increase once again.
6. Record the uncertainties in the collected data.

Data Collection and Processing:
- Collect appropriate data and present this in a way which will allow for easy interpretation.
- Include the uncertainties of measurement in your recorded data.
- Plot a suitable graph from which you can analyse the intensity and degree of polarisation of the polarized light in order to verify Malus' Law.
- Include the uncertainty in your calculated answer.

Conclusion and Evaluation:
- Make valid conclusions related to intensity of plane polarized light and Malus' Law
- Explain the results in terms of degree of polarization and angles of the crossed Polaroids
- Evaluate the method, including any modifications you had to make to overcome problems
- Include an evaluation of the apparatus used.
- Suggest ways in which the procedure could be modified in order to improve future investigations.

INVESTIGATING BREWSTER'S LAW

As light reflects from the surface of a non-metallic substance, it becomes partially plane-polarized and the degree of polarization depends upon the viewing angle of the reflected light. At a certain angle, the light will be completely plane polarized and this occurs at "Brewster's Angle" (named after the Scottish physicist, Sir David Brewster). It is possible to calculate the refractive index of the reflecting medium using Brewster's angle.

IB Criteria Assessed
Data Collection and Processing, Conclusion and Evaluation

Criteria	Aspect			Level
assessed	1	2	3	awarded
D				
DCP				
CE				

Aim:
1. To verify Brewster's Law for (partially) plane-polarized reflected light.
2. To use the data collected to find the refractive index of a piece of Acrylic.

Theory:
The relationship between refractive index and Brewster's angle can be given by $n = \tan\theta_B$

SAFETY FIRST – DO NOT LOOK DIRECTLY AT THE LASER LIGHT

Diagram:

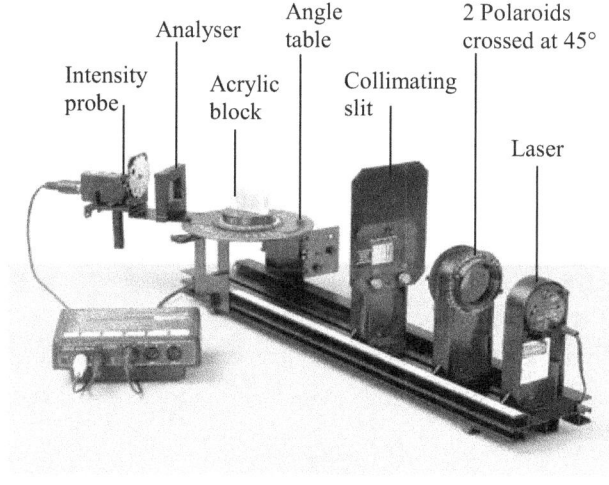

Intensity probe, Analyser, Acrylic block, Angle table, Collimating slit, 2 Polaroids crossed at 45°, Laser

Method:
1. Set up the apparatus as shown in the photograph. (The laser light is already polarised, so the 2 Polaroids crossed at 45° unpolarizes this laser beam).
2. Direct the laser beam onto the surface of the acrylic plastic block.
3. Set the angle of the analyser so that it is oriented horizontally, to eliminate the unpolarized reflected light.
4. Change the angle at which the laser beam is incident on the surface of the plastic and make necessary adjustments so that the reflected beam enters the intensity probe.
5. Record pairs of data for angle and intensity, until an angle is reached where the intensity of the reflected light falls to a minimum.
6. Continue past this minimum – the intensity of the reflected light should start to increase once again.
7. Record the uncertainties in the collected data.

Data Collection and Processing:
- Collect appropriate data and present this in an appropriate way, allowing for easy interpretation.
- Plot a suitable graph from which you can analyse the intensity and degree of polarisation of the reflected light for various angles of reflection.
- Use the graph to determine Brewster's angle and the refractive index for the Acrylic. Include the uncertainty in your calculated answers.

Conclusion and Evaluation:
- Quote your values found for Brewster's angle and refractive index for Acrylic and compare these to literature values.
- Explain the results in terms of degree of plane-polarization from a non-metallic reflector's surface.
- Evaluate the method, including any modifications you had to make to overcome problems. Include an evaluation of the apparatus used.
- Suggest ways in which the procedure could be modified in order to improve it for the future.

INVESTIGATING THE FOCAL LENGTH OF A CONVERGING LENS

Aim:
To find the focal length of a converging lens by experimental method

Method:
1. Set the apparatus up as shown in the diagram, so that the lengths u and v are equal and so that the image of the candle appears in focus on the screen.
2. Record lengths u and v.
3. Move the candle away from the lens by 20cm more and adjust the screen distance (v), until the image is one again in focus. Record u and v in the table.
4. Repeat step 3, three more times (moving the candle further away from the lens). Record u and v each time.
5. Place the candle and screen back in the original position found in step 1.
6. Move the screen away from the lens by 20cm more and adjust the candle distance (u), until the image is one again in focus. Record u and v.

IB Criteria Assessed
Data Collection and Processing, Conclusion and Evaluation

Criteria assessed	Aspect			Level awarded
	1	2	3	
D				
DCP				
CE				

Diagram:

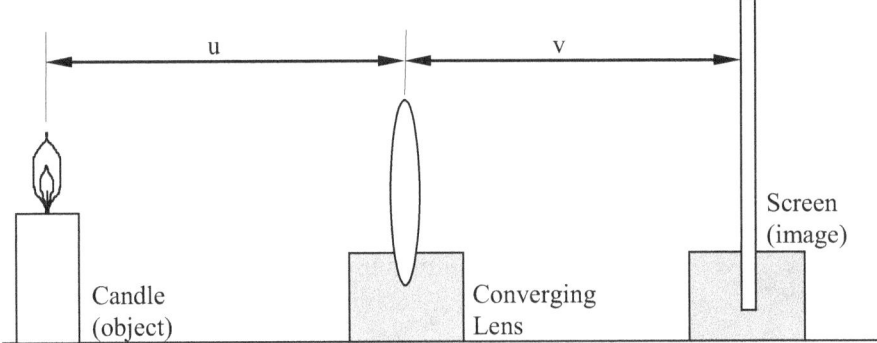

Theory:
The relationship between u, v and the focal length f for a converging lens is:- $\dfrac{1}{v} + \dfrac{1}{u} = \dfrac{1}{f}$

Data Collection and Processing:
- Collect and record pairs of data (u and v) including units and uncertainties.
- Present these data clearly in a suitable table.
- Process your raw data in a way which will allow you to accurately calculate the value of f (the focal length of the lens)
- Take into account any errors or uncertainties in your processed data.

Conclusion and Evaluation:
- Give a conclusion and explanation of your results; compare to quoted values if possible.
- Your explanation should include scale diagrams to explain some of the measurements taken. e.g.
    a. the object close to the lens (between pole and focal point)
    b. the object and image at the same position.
    c. the object far away from the lens.
- Evaluate the above procedure (method) and apparatus used, including limitations, weaknesses or errors.
- Suggest ways of improving the investigation.

INVESTIGATING MELDE'S EXPERIMENT

Aim:
In this experiment you will collect raw data on frequency and tension in a vibrating string and your task is to try and find some simple mathematical relationship between them. To collect data and, from it, produce an equation which we can use to make predictions about the behaviour of a system, is a very important process in science.

IB Criteria Assessed
Data Collection and Processing, Conclusion and Evaluation

| Criteria | Aspect | | | Level |
assessed	1	2	3	awarded
D				
DCP				
CE				

Apparatus:
Signal generator, vibrator, leads, 3 m of cotton, eight 50 g masses, mass hanger, pulley, ramp.

Diagram:

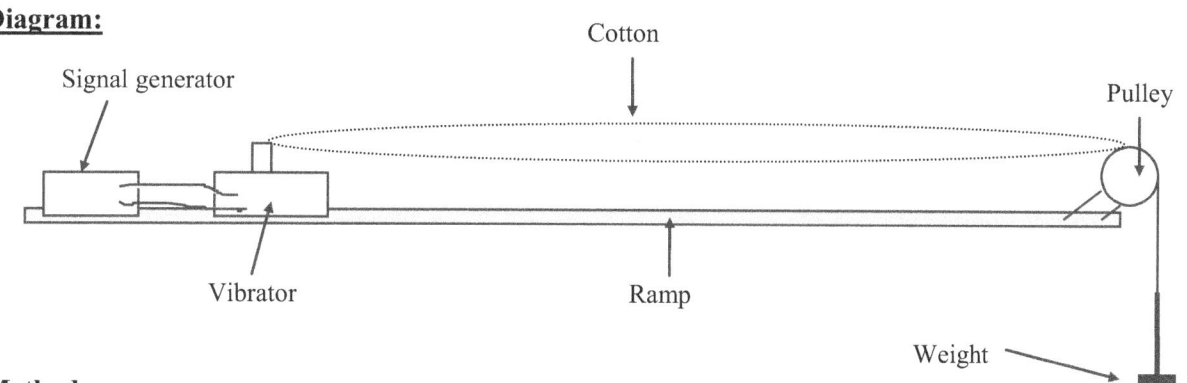

Method:
Put a 50 g mass on the string and adjust the signal generator until you get a large amplitude wave (as shown in the diagram). Note the frequency of the signal generator. Repeat this several times in steps of 50 g.

Theory:
There is clearly some sort of relationship between "frequency" and "tension", but what is it? It might be one of the following:

- frequency $\propto$ Tension in string
- frequency $\propto$ (Tension in string)^2
- frequency^2 $\propto$ Tension in string

Data Collection and Processing:
- Collect of pairs of results for f and T for when the string gives a large single amplitude
- Present these data (together with uncertainties) in an appropriate table.
- Plot a suitable graph which will enable you to determine the relationship between the fundamental frequency and the tension in the string

Conclusion and Evaluation:
- Evaluate the data that you have collected and analysed. What does the textbook say about the relationship between the fundamental frequency and the tension in the string? Compare your conclusion to this.
- Evaluate the procedure, including any modifications you had to make to overcome problems. Include an evaluation the apparatus used.
- Suggest ways in which the procedure could be modified in order to improve it for the future.

INVESTIGATING RESONANCE – DETERMINING THE VELOCITY OF SOUND

Aim:
Sound travels at approximately 330 m/s. This is too fast to measure with a stop watch. A clever way of measuring the speed of any wave (sound, light, radio etc.) is to freeze it. With a stationary wave, since it is not moving, it is easy to measure the wavelength and, if you also know the frequency of the wave, you can calculate the speed of the wave. This is yet another example of the use of a clever "trick" in order to measure something which is very difficult to measure directly.

IB Criteria Assessed
Data Collection and Processing, Conclusion and Evaluation

Criteria assessed	Aspect			Level awarded
	1	2	3	
D				
DCP				
CE				

Apparatus:
Large measuring cylinder, glass tube, set of tuning forks, metre rule.

Diagram:

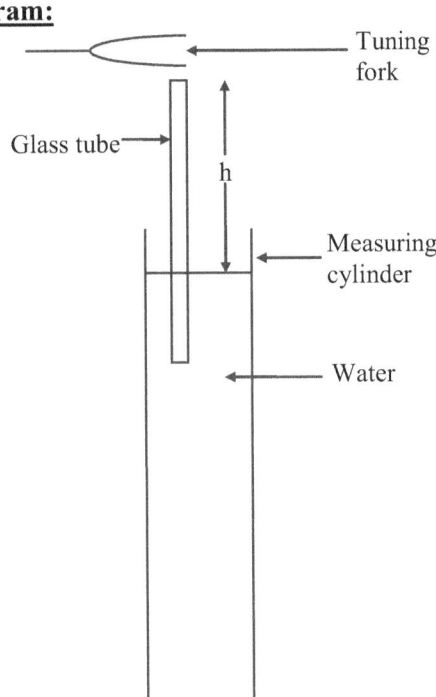

Tuning fork

Glass tube

h

Measuring cylinder

Water

Method:
1. Hold the vibrating tuning fork above the glass tube.
2. Move the tube up and down until you find the shortest length of the air column that produces a loud sound.
3. Measure length "h". Note the frequency of the tuning fork.
4. Repeat this for tuning forks of several different frequencies.

Theory:
You have found the fundamental resonance length for the different frequencies. The length at which resonance occurs for the different frequencies is a function of the wavelength of the sound wave in the glass tube. The relationship between the frequency, wavelength and speed of the wave is given by;

$$v = f \lambda$$

Data Collection and Processing:
- Collect of pairs of results for f and h for when the sound from the air column is at its loudest.
- Present these data (together with uncertainties) in an appropriate table.
- Plot a suitable graph from which you can calculate the speed of sound.
- If your graph does not go through the origin, then there may be a reason for this. (hint - look in your text book for "end correction")

Conclusion and Evaluation:
- Evaluate the data that you have collected and analysed. Compare the speed of sound obtained experimentally with the value published in your text-book.
- Evaluate the method, including any modifications you had to make to overcome problems. Include an evaluation the apparatus used.
- Suggest ways in which the procedure could be modified in order to improve it for the future.

DETERMINING THE WAVELENGTH OF LASER LIGHT USING YOUNG'S DOUBLE SLITS

Aim:

This experiment is historically very important because with it Thomas Young showed that light is a wave and so finished off Newton's "corpuscular theory" of light. If light is a wave then it should produce interference and also it should be possible to use the interference pattern to measure the wavelength of the light - this is exactly what Young was able to do - therefore proving that light is a wave.

Method:

Put the double slits in a stand and clamp as far as possible from the wall or screen. Aim the laser beam at the slits. Measure the distance from the slits to the wall and also the distance between the dots on the wall. Use the Vernier to measure the double slit separation.

IB Criteria Assessed

Data Collection and Processing, Conclusion and Evaluation

Criteria assessed	Aspect			Level awarded
	1	2	3	
D				
DCP				
CE				

DO NOT SHINE THE LASER LIGHT IN YOUR EYES - IT COULD BLIND YOU ! !

Apparatus:

Laser, Vernier gauge, metre rule, double slits, clamp and stand.

Diagram:

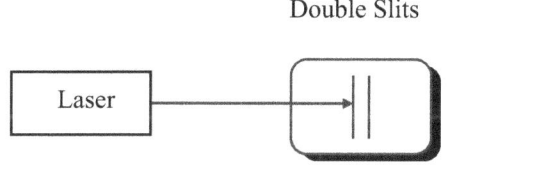
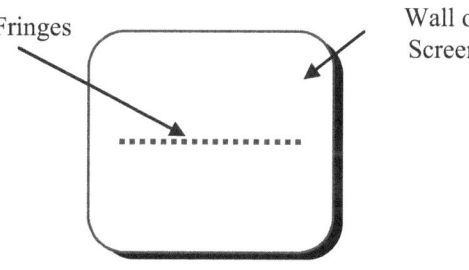

Theory:

Use the equation:

$$\text{Wavelength} = \frac{\text{Slit separation x Fringe width}}{\text{Distance from slits to screen}}$$

Data Collection and Processing:

- Show the results for the experiment in a suitable table. Include uncertainties.
- Calculate a value for the wavelength of laser light.
- Calculate the actual and percentage errors in this value, based on the uncertainties in the apparatus that you have used.

Conclusion and Evaluation:

- Evaluate the data that you have collected and analysed. The correct value for He / Ne laser light is given in the data book. Try to identify the errors in your experiment that caused the difference between your result and the true result.
- Evaluate the procedure, including any modifications you had to make to overcome problems. Include an evaluation the apparatus used.
- Suggest ways in which the procedure could be modified in order to improve it for the future.

DETERMINING THE WAVELENGTH OF LIGHT USING A DIFFRACTION GRATING

Aim:
To determine the wavelengths of red, green, and blue light by experimental method using a diffraction grating.

Method:
1. Place the lamp, lens and screen in the positions shown in the diagram.
2. Adjust the lens distance to form a focussed image on the screen.
3. Place the diffraction grating in the position shown.
4. Adjust the diffraction grating until clear fringes are formed on the screen.
5. Measure the length l and corresponding distance x and record them in a suitable table. (When using white light, three measurements can be made for x – from the central bright fringe to the red, green and blue parts of the first fringe).
6. Repeat for various distances of l and x.

IB Criteria Assessed
Data Collection and Processing
Conclusion and Evaluation

Criteria assessed	Aspect			Level awarded
	1	2	3	
D				
DCP				
CE				

Theory:

Wavelength n λ = d sin θ

where   d = diffraction grating slit width
         d = 1/N
         N = number of lines per mm

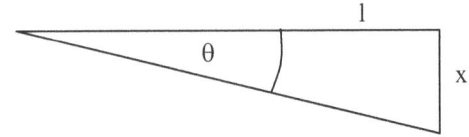

But for small angles of θ, tan $\theta \cong$ sin θ

$$\therefore \text{ Wavelength n } \lambda = \frac{d\,x}{l}$$

Diagram:

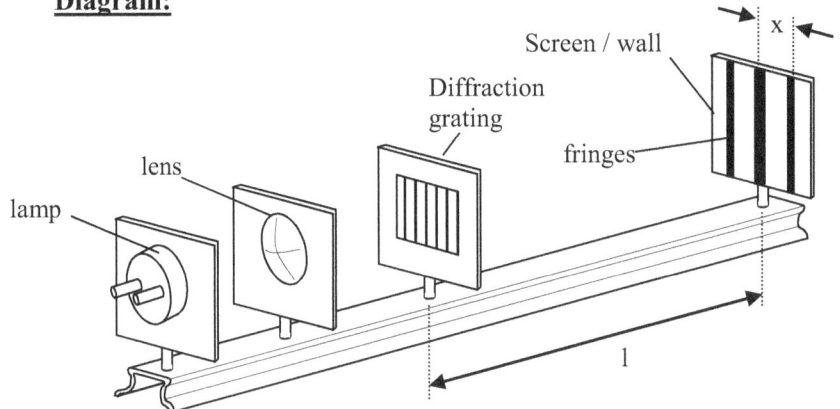

Screen / wall
Diffraction grating
lamp
lens
fringes
x
l

Data Collection and Processing:
- Collect of pairs of results for l and x for the colours red, green and blue.
- Present these (together with uncertainties) in an appropriate table.
- Plot a suitable graph which, combined with the theory above, will enable you to calculate the wavelengths of the three colours of light.

Conclusion and Evaluation:
- Evaluate your results.
- Compare your values for the wavelength of the 3 colours of light to those quoted in the text book.
- Comment on any possible sources of error, especially regarding the small angle approximation used in the theory above.
- Identify any weaknesses and suggest ways of improving the investigation.

INVESTIGATING THE POWER OF AN ELECTRIC HEATER

Aim:
1. To obtain a relationship between the power dissipated by an electric heater, its resistance and the current flowing.
2. To use appropriate mathematics to decide if your results prove or disprove the equation.

Apparatus:
50 cm of resistance wire, variable D.C. power supply, ammeter, voltmeter, thermometer, stopwatch, leads, 250 cm^3 beaker.

IB Criteria Assessed
Data Collection and Processing

Criteria assessed	Aspect			Level awarded
	1	2	3	
D				
DCP				
CE				

Diagram:

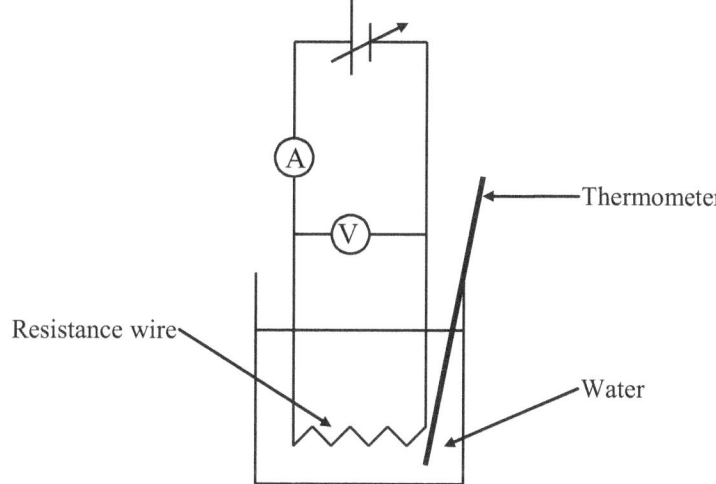

Method:
To produce the resistance coil, wrap the wire round a pencil. Set up the circuit. Take the temperature of the water. Set the current to 1 A, stir the water occasionally, and after 5 minutes take the new temperature Repeat the process at various currents.

Data Collection and Processing:
- Collect and record pairs of results for Voltage and Current including units and uncertainties.
- Present these data clearly.
- Process your raw data in a way which will allow you to accurately calculate the power received by the water.
- Take into account any errors or uncertainties in your processed data.

INVESTIGATING RESISTANCE WIRE

Copper and other good electrical conductors allow "free" (conduction) electrons to pass through them very easily. The molecular arrangement in a piece of Nichrome wire on the other hand disrupts this free flow of electrons and provides some resistance to their movement. There are many factors which determines the amount of resistance that the piece of wire provides.

Aim:
You are to investigate a factor that affects the resistance of a piece of Nichrome wire.

Apparatus:
One metre of Nichrome wire, plus any other materials you may need.

IB Criteria Assessed
Design,
Data Collection and Processing,
Conclusion and Evaluation

| Criteria | Aspect | | | Level |
assessed	1	2	3	awarded
D				
DCP				
CE				

Design:
Design a procedure that will allow you to investigate a factor (or factors) that affect the resistance of a piece of wire. This procedure should include the following sections:
- Defining the Problem and selecting variables:
- Controlling the Variables:
- Developing a method for collecting data:

Include a quantitative hypothesis for your investigation.

Data Collection and Processing:
- Show the results for the experiment in a suitable table. Include uncertainties.
- Use suitable graphs to allow for a full analysis to be carried out on your chosen variable(s)

Conclusion and Evaluation:
- Evaluate the data that you have collected and analysed. Make a suitable conclusion. Compare the result obtained from your experiment to literature values (if possible)
- Evaluate your own plan, including any modifications you had to make to overcome problems. Include an evaluation of the apparatus used.
- Suggest ways in which the procedure could be modified in order to improve it for the future.

DETERMINING ENERGY DENSITY OF FUELS

All fuels that can be burned have a particular energy density, or how much energy is packed in per kilogram of fuel. Usually, for something like cars, we want fuels that have a high energy density, so we don't have to lug so much around with us. One benefit of gasoline over coal is that it has a higher energy density. In this lab, you will compare the energy density of fuels that can be used easily in the laboratory.

IB Criteria Assessed

Data Collection and Processing, Conclusion and Evaluation

Criteria assessed	Aspect			Level awarded
	1	2	3	
D				
DCP				
CE				

Aim:
1. To compare the relative energy density of fuels.
2. To compare our measured energy densities with the actual literature values.

Theory:
The equation for energy density is….

Energy density = energy of fuel / mass of fuel

A rough assumption will be that the energy from burning the fuel can be put directly into heating water.

If so, energy to heat water is = $mc\Delta T$, where

m = mass water
c = specific heat capacity of water
ΔT = rise in temp. of water

Diagram:

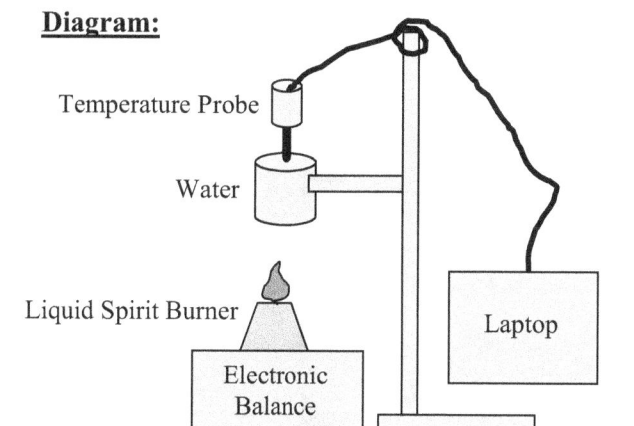

The mass of fuel used can be measured by the difference in mass of a spirit burner as it burns.

Method:
1. Set up the apparatus as shown in the diagram.
2. Use a reasonable amount of water given the size of your flame. 100 mL is a good starting point. Record this.
3. Record the initial temp of your water.
4. Lower the water (which should be on wire gauze) just over the flame.
5. Zero the scale so that it shows negative values as flame burns.
6. Set the laptop and temperature probe so that they measure water temperatures. Stir constantly.
7. Record appropriate data so as to obtain the energy density of water.
8. Record the uncertainties in the collected data.
9. Repeat this method using other fuels. Keep in mind that other fuels may heat the water much slower or faster, and you may want to use different amounts of water to allow for this.

Data Collection and Processing:
• Collect appropriate data and present this in a way which will allow for easy interpretation.
• Include the uncertainties of measurement in your recorded data.
• Plot a suitable graph from which you can analyse the energy density of the fuels.
• Include the uncertainty in your calculated answer.

Conclusion and Evaluation:
• Make valid conclusions related to fuel densities found. Compare the fuel densities relative to each other, and to literature values.
• Evaluate the method, including any modifications you had to make to overcome problems
• Include an evaluation of the apparatus used.
• Suggest ways in which the procedure (and apparratus) could be modified in order to improve future investigations.

INVESTIGATING LENZ'S LAW USING THE MOTION OF A FALLING MAGNET

Aim:
1. To find the motion of a magnet in free fall in a metal tube.
2. Use the data that you collected to Illustrate Lenz's and Faraday's Law.

IB Criteria Assessed
Data Collection and Processing, Conclusion and Evaluation

Criteria assessed	Aspect			Level awarded
	1	2	3	
D				
DCP				
CE				

Diagram:

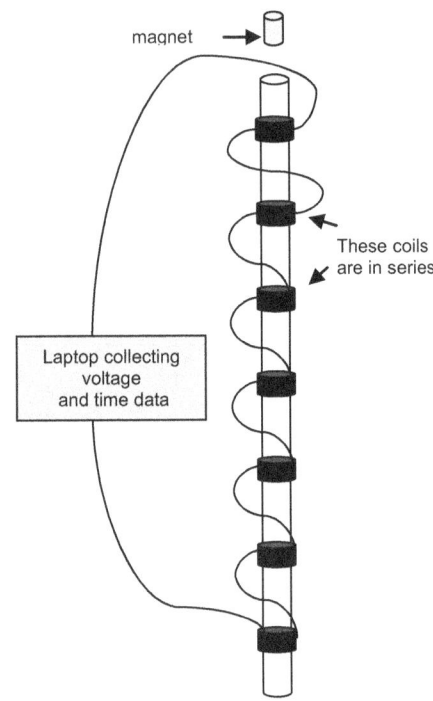

magnet →

These coils are in series

Laptop collecting voltage and time data

Method:
1. Obtain a long aluminium or copper tube, which has fixed coils attached to it.
2. Attach a wire from the top of the first coil to the Logger Pro interface box leads.
3. Run a second wire from the bottom of the first coil to the top of the second coil. Then run a wire from the bottom of the second coil to the top of the third coil. Continue this until you have all the coils connected in series.
4. Connect the bottom wire of the last coil to the 2nd lead from the Logger Pro.
5. Open the Logger Pro Software. Open the Experiments file. Open the voltage file.
6. Estimate the time that the magnet takes to fall through the tube. Add about 5 seconds to this and set this as your horizontal axis. Voltage will be your vertical axis.
7. Hit Collect, then drop the magnet in the pipe.
8. Measure the distance from the first coil to all the other coils
9. Construct a well labelled data table for you data.

Theory:
Lenz's Law and Faraday's Law both allow one to predict aspects of induced current when there is a changing magnetic field.

Data Collection and Processing:
- Collect appropriate data.
- Plot a suitable graph from which you can analyse the motion of the magnet as it falls through the tube.

Conclusion and Evaluation:
- Using your graph, analyse the type of motion of the falling magnet.
- Using your knowledge of the laws of electromagnetic induction, explain why this type of motion occurs.
- Evaluate the method, including any modifications you had to make to overcome problems. Include an evaluation of the apparatus used.
- Suggest ways in which the procedure could be modified in order to improve it for the future.

INVESTIGATING MAGNETS

You will probably have played around with magnets at one time or another. You may have noticed that a magnet will "stick" to some metals while it simply doesn't want to stick to others. You may have noticed that two magnets attract each other under certain circumstances while they push each other apart (repel) under a different arrangement.

Aim:
You are to investigate the force between two magnets

Apparatus:
Two small "magnadur" permanent magnets, plus any other materials you may need.

Planning:
Design a procedure that will allow you to investigate the factor (or factors) that affect the force between two magnets. This procedure should include the following sections:
- Defining the Problem and selecting variables:
- Controlling the Variables:
- Developing a method for collecting data:

Include a quantitative hypothesis for your investigation.

Data Collection and Processing:
- Show the results for the experiment in a suitable table. Include uncertainties.
- Use suitable graphs to allow for an analysis to be carried out on your chosen variable(s)

Conclusion and Evaluation:
- Evaluate the data that you have collected and analysed. Make a suitable conclusion.
- Evaluate your own plan, including any modifications you had to make to overcome problems. Include an evaluation of the apparatus used.
- Suggest ways in which the procedure could be modified in order to improve it for the future.

IB Criteria Assessed

Design,
Data Collection and Processing,
Conclusion and Evaluation

Criteria assessed	Aspect			Level awarded
	1	2	3	
D				
DCP				
CE				

VERIFYING THE EQUATION "F = B I L" USING A CURRENT BALANCE

Aim:
The 2 benefits of this experiment are:
- it is important to know how to carry out a controlled scientific experiment that will step by step isolate and test each of the variables in an equation - in this case, "B", "I" and "L". ("controlling the variables")
- this experiment needs a lot of careful manipulation of the equipment to make it work perfectly. This precision and patience is vital in scientific experiments.

IB Criteria Assessed
Data Collection and Processing
Conclusion and Evaluation

Criteria assessed	Aspect			Level awarded
	1	2	3	
D				
DCP				
CE				

Apparatus:
2 Magnadur magnets and holders, 1 m of thick copper wire, variable d.c. power supply, 2 metal pivots, crocodile clips, 50 cm of ticker tape.

Diagram:

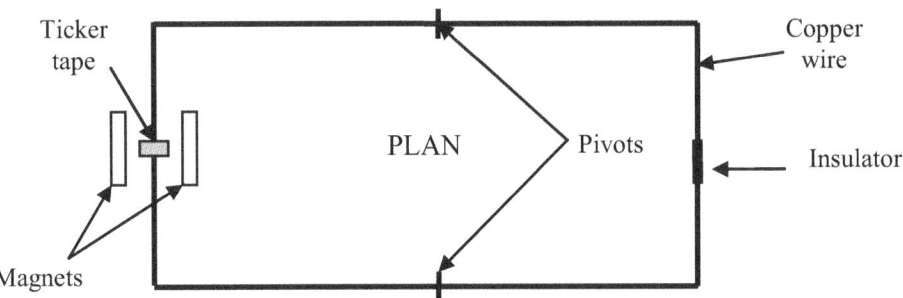

Method:
The copper frame is adjusted until it balances horizontally. Crocodile leads are connected to the 2 pivots and current passes through the left hand side of the copper frame (not the right hand side because the insulator blocks the current). The current direction is chosen so that the frame is pushed upwards by the electromagnetic force. A small piece of ticker tape is added so that the frame once again balances horizontally.

Theory:
The upward force (F) on the copper wire is given by the equation, F = B x I x L.

Data Collection and Processing:
- Collect and record pairs of results for Force and Current including units and uncertainties.
- Present these data clearly.
- Process your raw data in a way which will allow you to accurately calculate the value of the magnetic field strength, B.
- Take into account any errors or uncertainties in your processed data.

Conclusion and Evaluation:
- Give a conclusion and explanation of your results; compare to literature values if possible.
- Evaluate the above procedure (method) and apparatus used, including limitations, weaknesses or errors.
- Identify any weaknesses and suggest ways of improving the investigation.

INVESTIGATING ELECTROMAGNETS

Aim:

All scientists, before they start any experiment, will attempt to "guess" the result, or make a hypothesis about the experiment. This shapes how and what they will do in their experiment. In this practical, before you start, you will write down your predictions of the results. It is best, if you can, to make your predictions quantitative - that is to say use numbers, including an explanation.

This process is very important in science. A good scientist can save a lot of time and money by using their experience and intuition to predict what they think will be useful experiments - instead of going down a dead end and getting nowhere.

IB Criteria Assessed

Design,
Data Collection and Processing,
Conclusion and Evaluation

Criteria assessed	Aspect			Level awarded
	1	2	3	
D				
DCP				
CE				

Apparatus:

An iron nail plus any other materials you may need.

Planning:

Design a procedure that will allow you to investigate the factor (or factors) that affect the strength of an electromagnet. This procedure should include the following sections:

- Defining the Problem and selecting variables:
- Controlling the Variables:
- Developing a method for collecting data:

Include a quantitative hypothesis for your investigation.

Data Collection:

- Show the results for the experiment in a suitable table. Include uncertainties.

Data Processing and Presentation:

- Use suitable graphs to allow for an analysis to be carried out on your chosen variable(s)
- Include uncertainties in your processed data

Conclusion and Evaluation:

- Evaluate the data that you have collected and analysed.
- Evaluate your own plan, including any modifications you had to make to overcome problems. Include an evaluation of the apparatus used.
- Suggest ways in which the procedure could be modified in order to improve it for the future.

INVESTIGATING ELECTROMAGNETIC INDUCTION

Aim:
From what we have learned through Faraday's law, we know that a changing magnetic field can induce a current in a loop of wire. Several factors can affect the strength of this induced current.

Apparatus:
- Wire
- Bar Magnets
- Galvanometer
- Anything else that could reasonably be found in a normal physics classroom.

Design:
Design a procedure to test how a certain variable (of your choice) may affect the strength of the induced current. This procedure should include the following sections.

- Defining the Problem and selecting variables:
- Controlling the Variables:
- Developing a method for collecting data:

Also include a hypothesis and a sketch graph of what you think will happen.

IB Criteria Assessed
Design

Criteria assessed	Aspect			Level awarded
	1	2	3	
D				
DCP				
CE				

INVESTIGATING THE EFFICIENCY OF A TRANSFORMER

Aim:
An engineer is often employed in order to test the efficiency of a machine under a variety of conditions. It could be that the machine is highly efficient over a narrow range but outside of this range, the efficiency drops drastically. It is important for a manufacturer or user to know what this range is. This is what you are going to do here. The aim is to produce an efficiency graph for a transformer.

IB Criteria Assessed
Data Collection and Processing

| Criteria | Aspect | | | Level |
assessed	1	2	3	awarded
D				
DCP				
CE				

Method:
1. Connect up the circuit and set the power supply to 2 V a.c.
2. Measure the current and voltage on the primary and secondary sides of the transformer (you will have to disconnect and reconnect the meters as you only have 2).
3. Repeat the process for several different voltages up to 12 V.

Apparatus:
Demountable transformer, leads, 2 multimeters, variable a.c. power supply, 12 V light bulb.

Diagram:

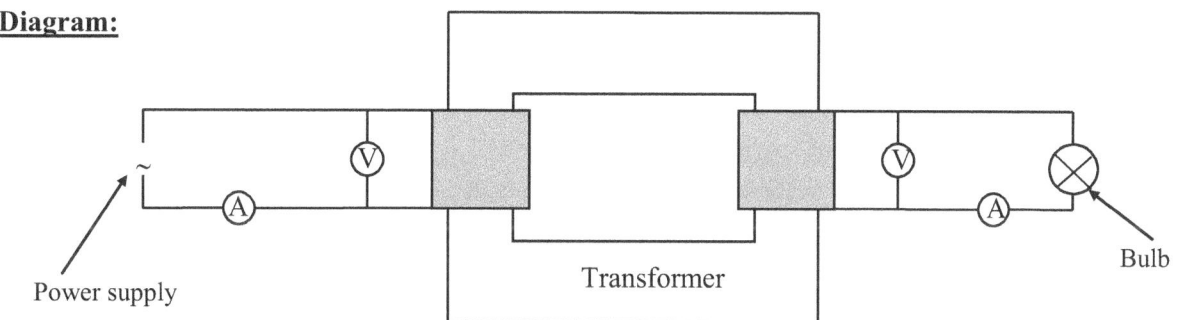

Power supply   Transformer   Bulb

Data Collection and Processing:
- Collect data for current and voltage on the primary and secondary side of the transformer.
- Present this data in a suitable table. Include uncertainties.
- Process your raw data in a way which will allow you to accurately calculate the efficiency
- Take into account any errors or uncertainties in your processed data.
- Plot a suitable graph which will allow a relationship between efficiency and voltage to be analysed. Comment on this relationship.

DETERMINING THE EMF AND INTERNAL RESISTANCE OF A CELL

Aim:
To determine the emf and internal resistance of a cell (battery) using a graphical method.

Method:
1. Set up the circuit as shown in the diagram with the variable resistor set to its maximum value.
2. Before closing the switch, record the reading, V, on the voltmeter
3. Close the switch
4. By adjusting the variable resistor, obtain pairs of voltmeter and ammeter readings over the widest possible range.
5. Open the switch after each pair of readings and only close it for as long as is necessary to obtain each pair of readings.

Apparatus:
Dry cell (battery), ammeter, voltmeter, variable resistor, switch.

Diagram:

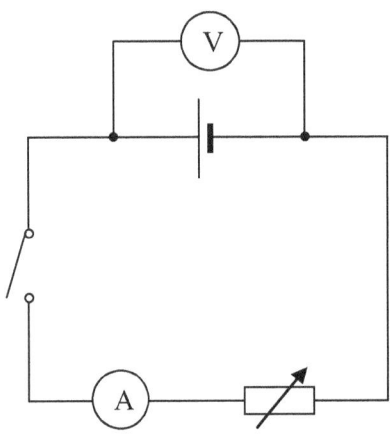

IB Criteria Assessed
Data Collection and Processing
Conclusion and Evaluation

Criteria	Aspect			Level
assessed	1	2	3	awarded
D				
DCP				
CE				

Data Collection and Processing:
- Collect and record pairs of results for Voltage and Current including units and uncertainties.
- Present these data clearly.
- Process your raw data in a way which will allow you to accurately calculate the value of emf and internal resistance for the cell.
- Take into account any errors or uncertainties in your processed data.

Conclusion and Evaluation:
- Give a conclusion and explanation of your results; compare to literature values if possible.
- Evaluate the above procedure (method) and apparatus used, including limitations, weaknesses or errors.
- Identify any weaknesses and suggest ways of improving the investigation.

INVESTIGATING RADIOACTIVE DECAY (SIMULATION USING DICE)

Aim:
This experiment simulates radioactive decay. Instead of nuclei decaying, we will be looking at dice with the number "1" showing. These dice will represent decayed atoms and will be removed at each step.

Apparatus:
100 dice, large container.

Method:
1. Put the 100 dice into a container.
2. Shake the container and pour the dice onto the desk.
3. Remove decayed nuclei and record the number involved.
4. Put the remaining dice back into the container and repeat steps 2 and 3 several times.
5. Repeat, to get two full sets of data.
6. Share your results with the other teams so that you can calculate an average value for the "number of remaining dice" for each roll.

IB Criteria Assessed
Data Collection and Processing, Conclusion and Evaluation

Criteria assessed	Aspect			Level awarded
	1	2	3	
D				
DCP				
CE				

Data Collection and Processing: (BOTH Standard and Higher level)
-   Record your results for the simulation in a suitable table.
-   Your results table and the presentation of data should include any uncertainties associated with the apparatus that you have used.
-   Draw a suitable graph that will allow you to calculate the half-life of your decay model.
-   Show your working clearly and explain your method used.
-   State your final value for the half-life for your decay model
-   Explain if the background count rate was included in this graph.

Data Collection and Processing: (Higher level only)
-   The radioactive equation can be written:

$$N = N_O e^{-\lambda t}$$

-   Rearrange this formula so that it is of the form y = mx + c
-   Draw a suitable log graph that will allow you to obtain a value for the decay constant, λ and calculate the half-life, t, of your decay model.
-   Compare the two values for half-life obtained from the graphs.
-   Which method do you think is more accurate? State your reasons.

APPENDICES

APPENDIX 1 – FORM 4/PSOW

Submit to: _____   Arrival date: _____   Session: _____

School code: _____   School name: _____

Subject: _____   Level: _____

Candidate name: _____   Candidate number: _____

Date(s)	Outline of experiments / investigations / projects (include title and a brief description)	ICT	Topic / option	Time (hrs)	Levels awarded D	DCP	CE
	Investigating Murphy's Law						
	Investigating the Fall of a Coffee Filter						
	Investigating the Simple Pendulum						
	Investigating Uncertainties – Measuring Instrument Circus						
	Investigating Errors and Uncertainties in Experiments						
	Determining Stiffness of Steel by the Oscillations of a Hacksaw Blade						
	Investigating the Stopping Distance of a Bicycle						
	Investigating the Torsional Pendulum						
	Investigating Projectiles						
	Investigating Forces in Equilibrium						
	Investigating Error and uncertainties using Acceleration due to gravity						
	Investigating the Flight of an Elastic Band						
	Investigating Work Done and Energy Transferred on an Inclined Plane						
	Investigating the Ballistic Pendulum						
	Investigating Energy Transfer and Energy Loss of a Rolling Ball						
	Investigating Newton's 2nd Law of Motion using a Ticker timer						
	Investigating Hooke's Law						
	Investigating Springs in Series and Parallel						
	Investigating Circular Motion						
	Investigating Ping - Pong Balls						
	Determining Specific Heat Capacity by The Electrical Method						
	Determining Specific Heat Capacity by The Method of Mixtures						
	Investigating Specific Latent Heat of Vaporisation of Water						
	Investigating Specific Latent heat of Fusion of Ice						
	Investigating the Power and Temperature of the Sun						
	Investigating Charles' Law and Absolute Zero						
	Investigating Rate of Cooling						

	Investigating the Pressure / Volume relationship for a Balloon						
	Determining the Temperature of a Wire by Expansivity						
	Determining the Refractive Index of Glass by Real and Apparent Depth						
	Investigating Refraction of Light						
	Investigating Malus' Law						
	Investigating Brewster's Law						
	Investigating the Focal Length of a Converging Lens						
	Investigating Melde's Experiment						
	Investigating Resonance – Determining the Velocity of Sound						
	Determining the Wavelength of Laser Light using Young's Double Slits						
	Determining the Wavelength of light using a Diffraction Grating						
	Investigating the Power of an Electric Heater						
	Investigating Resistance Wire						
	Determining Energy Density of Fuels						
	Investigating Lenz's Law using the Motion of a Falling Magnet						
	Investigating Magnets						
	Verifying the Equation "F = B I L" Using a Current Balance						
	Investigating Electromagnets						
	Investigating the Efficiency of a Transformer						
	Investigating Electromagnetic Induction						
	Determining the Emf and Internal Resistance of a cell						
	Investigating Radioactive Decay (Simulation using Dice)						
	Group 4 Project						

Group 4 Project mark for Personal Skills (PS)
(same mark for students doing 2 subjects) /6

Summative Mark for Manipulative Skills (MS) /6

Two highest levels achieved

/6	/6	/6
/6	/6	/6

Total

/48

This total must also be entered on IBIS

For completion by the examiners			
Moderator	/6	/6	/6
	/6	/6	/6

Senior Moderator	/6	/6	/6
	/6	/6	/6

To be completed by the teacher:

Name: _____   Signature: _____   Date: _____

Candidate declaration: I confirm that this work is my own work and is the final version. I have acknowledged each use of the words or ideas of another person, whether written, oral or visual.

Candidate's signature: _____   Date: _____

APPENDIX 2 – THE GROUP 4 PROJECT

The following is an outline of one of the Group 4 Projects that I undertook with a group of Grade 11 students a few years ago while teaching at The United Nations International School, Hanoi, Vietnam. The following schedule and student sheets were put together into a booklet for each student and is just one way that the Group 4 project can be conducted. I happen to like this method of setting specific deadlines for the task and having the whole project performed over a two day period (ideally at the end of the first year of the IB course)

Science IB Group 4 Project 2006 - Introduction

Guidelines

The Group 4 project allows students to appreciate the environmental, social and ethical implications of science. It may also allow them to understand the limitations of scientific study. The emphasis is on interdisciplinary cooperation and the processes involved in scientific investigation. The exercise should be a collaborative experience where concepts and perceptions from across the science disciplines are shared. The intention is that students analyse a topic or a problem which can be investigated in each of the science areas.

The project should take between 10 – 12 hours, and can be divided up into the following stages:

Planning   Identifying topics to investigate from the perspective of your science discipline(s) under the central theme of "Science at UNIS".

Definition of Activities   Deciding who is going to do what. Complete the planning sheet.

Action   Carry out the investigation, either by research or practical work. Share your findings with the other members of your group (and with students of other groups if appropriate). Make a PowerPoint presentation of your research.

Evaluate   Discuss with your peers the strengths and weaknesses of your project. Complete the evaluation form.

Note: Students taking two sciences must produce a presentation which covers both sciences and evaluate the project from the perspective of both sciences.

Timing

Thursday June 8th   1.55pm. Planning and Definition of Activities stages. IB Group 4 guidelines distributed to all students in order to familiarise themselves with what the project intends to do and achieve – what the IB is looking for. Students will be organised into groups. Group brainstorming session follows. Provisional title and outline decided on by all students.

Friday June 9th   8.10am. Collect completed planning sheet (including apparatus list) from all students.

Friday June 9th   8.10am – 3.05pm Action stage. Data collection and initial data analysis on and around the UNIS campus.

Wednesday June 14th   8.10am – 11.50am Action stage. Data collection and initial data analysis on and around the UNIS campus.

Wednesday June 14th   12.35pm. Presentation of work. All students are requested to prepare a PowerPoint presentation of their research. These presentations must be emailed to Mr. Dickinson by no later than 12 noon of Wednesday June 14th. **No presentation will be accepted at the start of the lesson.**

Science IB Group 4 Project – Questions and Answers

1. *What is it?*
    Approximately 10 hours of research into an area decided by the students in that group and approved by a Science teacher.

2. *Does that mean that if I do two Sciences, I have to do two lots of 10 hours?*
    No, but you do have to investigate your area of research from the perspective of both of your science disciplines.

3. *How are the groups made up?*
    They are groups of 4 or 5 students from a mixture Science disciplines. For example, a group may have a Chemist, a Physicist, a Biologist and a couple of Environmental Systems students. The groups are decided by the teachers, not the students!

4. *When does it take place?*
    During the last two weeks of grade 11. During the initial meeting, the groups will have to decide on their area of study. They will then have to assign tasks to each student that will take about 10 hours of research. These plans should be submitted on the form provided to the teacher that group is assigned to for approval and/or advice and modification. One and a half days will be put aside for students to do their research. This will take place on **Friday 9th June** from 8:10 a.m. until about 3:05 and **Wednesday 14th June** from 8:10 a.m. until about 11.50.

5. *What happens after that?*
    Once each team member has collected all the data, the team will have to meet to decide how they will organise their presentation. Each team will have to prepare a presentation of 10-15 minutes showing what they did and what they found out. This will take place on **Wednesday 14th June**, during the afternoon (periods 4 and 5).

6. *What form should the presentation take?*
    Each team should prepare a PowerPoint presentation, which will be a permanent record of their research. This may contain photographs, graphs and other data. Every member of the team must take part in the presentation. You will have to answers from the audience, which will include students and teachers.

7. *What will I have to hand in for assessment?*
    You need to hand in a report on your part of the project **for each Science studied** to the teacher of that subject. A good way to do this is to produce a miniature of your PowerPoint presentation (9 slides per page). Remember to include your planning and evaluation sheets. This report will form a vital part of your science practical scheme of work (4PSOW) so **don't lose anything**.

8. *What skills will be assessed?*
    The IB Group 4 Project may only be assessed for Personal Skills. Make sure that you have familiarized yourself with the rubric for this criteria because the mark that you are awarded will count towards your final IB grade.

9. *What is the topic for research?*
    Science on and around the UNIS Hanoi, Tay Ho Campus.

Science IB Group 4 Project 2006 – Planning Sheet

Name ....   **Teacher's Name** ....

Provisional title of project

....

....

....

....

Outline of project

....

....

....

....

....

....

....

....

....

....

....

Apparatus list

....

....

....

....

....

....

....

....

Group members and science disciplines

Physics..

Biology..

Chemistry..

Environmental Systems..

Science IB Group 4 Project 2006 – Planning Sheet (continued)

Group member	Data Collection Responsibilities (One set for each Science subject)
	....
	....
	....
	....
	....

Teacher Comments

....
....
....
....
....
....
....
....
....
....

Science IB Group 4 Project 2006 – Evaluation Sheet

Name ....   Teacher's Name ....

Final title of project

....

....

....

....

Group members and science disciplines

Physics...

Biology...

Chemistry...

Environmental Systems..

Evaluation

Strengths: ....

....

....

....

....

....

Weaknesses: ....

....

....

....

....

....

Summary of Conclusions

....

....

....

....

....

....

....

....

Science IB Group 4 Project 2006 –Summary

Each year, the grade 11 science students take part in a cross-discipline science project as part of the IB Diploma Practical Scheme of Work. Science is Group 4 on the IB Diploma hexagon and the "Group 4 Project" is an opportunity for students to work with areas of science that are often new to them.

Under the umbrella title "Science at UNIS", the 35 grade 11 students worked in groups of 4/5 investigating areas of the UNIS Tay Ho campus from the perspective of all 4 science disciplines. Each group comprised a Physicist, a Biologist, a Chemist and a student studying Environmental Systems.

The 2½ day (15 hour) project allowed for the collection and analysis of data from the four corners of the UNIS campus. Projects included:-

The Pond
- Oxygen and pH levels
- Algae content
- Are the fish happy?
- Natural vs. forced oxygenation
- Sunlight and shade

Air-conditioning
- How does it work?
- Effects on the body
- Environmental impact of CFCs released by the refrigerant
- Efficiency

Coffee
- Effects of caffeine on heart-rate
- Coffee growing regions of the world (with a focus on Vietnam)
- Paper vs. Styrofoam – avoiding MacDonald law-suits
- The half-life of caffeine in the body.

The Swimming Pool
- Safe levels of chlorination
- Disposing of chlorine safely
- What should the water temperature be for a comfortable swimming experience?

Litter
- Bacteria in the litter-bins
- Which rubbish can be burnt to generate electricity?

The Canteen
- Energy content in food
- Bacteria in the canteen
- Nutritional content of food Efficiency of microwave cooking

The students were enthusiastic and enjoyed the experience of taking part in a science project that that was fun without the pressure of extensive write-up and evaluation.

APPENDIX 3 – POSSIBLE TOPICS FOR INVESTIGATION

While on my Level 1 workshop in Cali, Colombia 1997, Brian Seve gave me this list of titles for potential investigations – either to be added to the 4/PSOW, for possible ideas on topics or subtopics for the school's Group 4 Project or perhaps when a student is interested in doing their Extended Essay in Physics but can't think of a theme or a title.

- The shutter speeds of a camera
- The accuracy of aim of an air rifle, catapult, or improvised gun
- The true path of a ball thrown in air
- Water drops failing on water (flash pictures?)
- Splashing of moving drops hitting solids
- A narrow water trough as an accelerometer
- The profile of a rotating water surface
- The precession of a gyroscope
- Comparisons of human reaction times (between individuals; for different stimuli)
- Time taken by a switch to make or break contact
- Bouncing of relay contacts
- How much does the air pressure in a football matter?
- The performance of a firework rocket
- The bounce-time of a ball
- Factors affecting the friction of steel on ire
- The effect of oil films between sliding metal surfaces
- Does water absorb ultra-violet light?
- How long does the flash from a flash bulb last?
- How long does the flash from a xenon stroboscope last?
- How does the light coming through a slotted wheel stroboscope vary with time?
- Study the motion of a ball rolling on a turntable
- What does an air track collision look like from a moving point of view? (Moving camera)
- The distribution of speed, or of energy, among balls rolling randomly in a shaking tray
- The possible orbits of a pendulum bob
- The motion of the lip of a vibrating wire
- The performance of a water pump
- The performance of a fan
- The thrust of a propeller (in air, or in water)
- The energy delivered by a catapult
- Load and speed variations of a model aero-engine
- The fuel consumption of a model aero-engine
- The temperature changes and cooling of a model aero-engine
- The air supply to a model aero-engine
- Reduction of noise from a model aero-engine
- Factors affecting beam bending
- Factors affecting the buckling of a beam under compression

- Factors affecting the flexing of a rotating shaft
- The strength of girders of different construction (use balsa wood)
- The energy stored in a spiral clock spring
- Factors affecting the design of a good paddle wheel
- Making strong concrete bars
- The fracture of concrete by impact forces
- Effects of reinforcement on concrete
- The strength of fibreglass repairs (commercial fibreglass kits)
- Ice is said to be made less brittle by freezing sawdust into it. Is it?
- Variation of flow behaviour with rate of strain (silicone putty)
- Effects of heat-treating razor blades
- Heat-treatment of steel
- Heat-treatment of copper
- Heat-treatment of glass
- Flow patterns in glycerine (see Shapiro, A. H. Science Study Series, Shape and flow, Heinemann)
- Perspex is said to 'remember' that it has been deformed, for a while. Does it?
- The strength of human hair
- The strength of paper
- The properties of glued joints
- Making long lasting soap films
- Adhesion of glues to metals, fabrics, etc.
- How finely woven must umbrella material be?
- The changes in melting point of a solder with composition
- What is necessary for solder to flow?
- The strength of a soldered joint
- The bouncing of steel bails on glass
- Impact cracks when steel balls are dropped an glass
- Dents made in metals by balls pressed on them (Brinel hardness test)
- The heating and cooling of stretched rubber
- The creep of stretched rubber
- The strength and fracture of taut rubber bands
- The effect of temperature on stretched rubber
- Changes of length of hair with moisture content
- Factors affecting the growth of crystals
- The sagging of taut wires loaded in the middle
- The shape of a suspended loose chain

- Will a hole at the end of a crack help to stop the crack from spreading?
- What factors influence the production of good, uniform bubble rafts?
- The effect of various sorts of perforations on tearing paper
- The pressure-volume relation for a rubber balloon
- The effect of temperature changes on the flow of motor oils
- The design of a flow meter
- Reduction in pressure with fast flow (Bernoulli effect)
- Calibration of a V-slot flow meter (rate of flow from height of water in a V shaped slot)
- The drag on spheres and other shapes in an air stream
- The resistance to water flow of various plumbers' fittings (pipe, bends, etc.)
- The drag on objects towed in water (changes with length, depth of water, and many other factors)
- When does water flow in a tube become turbulent?
- The effect of changing the size or shape of the wings of a glider
- The penetration of projectiles into soft materials
- Load and speed variations for a parachute
- A water-driven rocket
- Measuring the viscosity of air
- Factors affecting the performance of an air track vehicle
- Making very big drops (oil in water and alcohol mixtures)
- How do Plateau spherules form?
- Soap films formed on spirals and other wire shapes
- The behaviour of bubbles rising in liquids
- The noise made by a kettle just before it boils (singing)
- The airflow in a room with a heater
- Smoke rings (a box with a hole at one end, and a flexible diaphragm at the other)
- Vortex rings in water (drop coloured water drops onto clear water)
- How much can a container be overfilled with water?
- How does water drip from a narrow jet?
- Variations in damping of a pendulum in air
- Water from a tap running into a flat basin sometimes forms a smooth ring of water,
- with a circular edge beyond which the flow is rougher. What decides the size of the ring?
- Where does dust collect? Why?
- Stiff standing rods will oscillate in an airflow. Investigate.
- The supporting of a ball on a jet of air, or of water
- The behaviour of coupled oscillators
- How much damping is needed to stop oscillations?
- Variable damping of a galvanometer
- Oscillations of drops
- Oscillations of rubber sheets

- Oscillations of soap films
- Oscillations of metal discs
- Oscillations of thin panels (e.g. doors, sheets of hardboard, sheets of metal)
- Oscillations of wire rings
- Oscillations of solid bars (notes from a xylophone)
- The factors affecting the performance of a sensitive flame
- How long does a sound last in a large hall?
- The propagation of sound at low pressures
- Can the motion of air in a sound wave be made visible?
- Slopping modes of oscillation in tanks of water
- How to isolate laboratory apparatus from vibrations
- 'Pearls in air'- what makes them form easily? (See Nuffield O-level Physics, Guide to experiments IV, experiment 21b)
- The resonance of a 'ticket timer'
- The frequency characteristics of a cheap gramophone pick-up
- The frequency response of a one-transistor amplifier with feedback
- Photographing waves on strings or springs
- The wakes of boats
- Waves in moving water
- Speed of waves in shallow water
- Breaking of waves
- The speed of ripples on water
- What are the shadows of waves on a ripple tank shadows of?
- The directional properties of a television aerial
- Variation in response of a dipole with length of the dipole
- Frequency range of a microphone
- Audible range of humans and animals
- The diffraction of sound waves
- Producing and detecting ultrasonic waves
- The pressure changes in the sound from an explosion
- Reflection or absorption properties of materials for microwaves
- Reflection or absorption properties of materials for sound waves
- Sound-absorbing tiles sometimes have perforated hardboard over an absorbent layer. Does the hole size matter?
- The behaviour of a loudspeaker cabinet at low frequencies
- The penetration of sound through double glazed panels
- Waves in circular dishes
- How good is a wax lens for microwaves?
- The colours of thin films of oil on water
- 'Shadows' of hot air from flames or heated objects
- The field of view of a simple telescope
- The depth of focus of a simple telescope

- The depth of focus of a microscope
- The resolution of a microscope
- Depth of focus of a camera
- Photography through a microscope
- Patterns in stressed materials between crossed polaroids
- Moiré fringes (Patterns from overlapped regular grids)
- Detection of small motions by interference methods (thermal expansion, compressibility)
- How much light is reflected at various angles by glass?
- The sensitivity of Kodak PI 53 paper at various wavelengths
- The adaptation to dark of the human eye
- The resolution of close-spaced objects by the eye
- Does photographic film fog equally if the light is bright and the exposure short, or if the light is dim and the exposure long?
- How big are the grains in a photograph ?
- How fast must a flicker be before it stops being observable?
- Make a diffraction grating by photographic reduction, and test it
- Do people vary in the range of wavelengths they can see?
- How quickly does the iris of the eye contract when the light is made brighter?
- Does the resolution of the eye depend on the illumination?
- The performance of a pin hole camera
- How much is scattered light polarised?
- A dynamo as a speedometer (conversion to accelerometer?)
- The efficiency of a dynamo
- The efficiency of an electric motor
- Load and speed variations of an electric motor
- Efficiency of a transformer
- Saturation effects in a transformer
- Effect of air gaps in transformers or electromagnets
- Eddy current losses in transformers (solid core)
- Stray fields around transformers
- The time taken for a fuse to blow
- The conduction of electricity by pencil lines on paper
- Conducting paper as a model for electric potential variations
- Potential variations in a tank of conducting liquid
- The time taken for ions to recombine (e.g. blown down-wind of a flame)
- How good are 10 per cent radio resistors?
- How good are 20 per cent radio capacitors?
- Torque-speed variations of a gramophone motor
- Energy emitted by a lamp bulb
- Lifetime of torch bulbs

- Does a photo-transistor respond instantly?
- Variations of resistance with strain
- How sensitive can a Wheatstone bridge be made?
- Resistance changes of human beings with variations in emotional state
- The running down and recovery of a dry cell
- How much charge can a home-made accumulator store?
- Electrolytic capacitors are said not to lose all their charge if short-circuited after being charged for some time. Is it so?
- An electroscope as a voltmeter
- The sensitivity of an electroscope as a charge measuring device Moving coil
- Milliammeters as ballistic galvanometers
- Make a capacitor microphone
- The variation of the field of a small coil with angle
- The contraction of a spiral carrying a current
- The effect of thickness of metal on eddy current forces
- How high will a 'jumping ring' jump? (A ring over an iron core with a coil carrying a.c. on the core)
- Frequency dependence of the impedance of an iron-cored inductor
- The dependence of the speed of a d.c. motor on field current
- Change in length of a nickel rod in a magnetic field
- The voltage from a thermocouple
- Temperature variations of transistor currents
- Is it true that a dry cell is the most expensive way to buy electricity?
- The design of an alternating current ammeter
- Behaviour of two LC circuits coupled together
- The design of an electronic exposure timer
- The energy balance of a photocell
- Electrical noise in a hot resistor
- Does a flame conduct electricity?
- Does hot air conduct electricity?
- What factors make for good deposits of copper in electrolysis?
- What factors affect heating by eddy currents?
- How does the resistance between two points on a conducting sheet vary with distance?
- How does the resistance between two flat plates in a tank of conducting liquid vary with their spacing?
- Make an electrostatic dust collector
- Magnesium oxide smoke collects in long fibres on electrodes at high potentials. Investigate. (Exclude draughts.)
- How does the resistance in an LC circuit affect the resonance?
- How does the electron current in a radio valve vary with filament temperature

www.ingramcontent.com/pod-product-compliance
Lightning Source LLC
Chambersburg PA
CBHW081549170526

45166CB00009B/2633